D0318542

Plate I (*Frontispiece*)

High Force on the River Tees. The lip of the fall is the Whin Sill; the Tynebottom
Limestone crops out at the foot

NATURAL ENVIRONMENT RESEARCH COUNCIL
INSTITUTE OF GEOLOGICAL SCIENCES

British Regional Geology

Northern England

(FOURTH EDITION)

By B. J. Taylor, B.Sc., I. C. Burgess, B.Sc.,
D. H. Land, B.Sc., D. A. C. Mills, B.Sc.,
D. B. Smith, B.Sc. and P. T. Warren, M.A., Ph.D.

Based on previous editions by
T. Eastwood

LONDON: HER MAJESTY'S STATIONERY OFFICE 1971

*The Institute of Geological Sciences
was formed by the incorporation of the
Geological Survey of Great Britain
and the Museum of Practical Geology
with Overseas Geological Surveys
and is a constituent body of the
Natural Environment Research Council*

© *Crown copyright 1971*

First published 1935
Fourth edition 1971

SBN 11 880141 4

Foreword to Fourth Edition

The mineral resources of northern England began to be exploited soon after the Roman Conquest, and the rocks of the region have been studied and written about since the earliest days of organized geology. The compilation of a handbook on the region therefore presents a problem of generalization without losing the main historical thread or neglecting any important feature of the rocks. This was admirably handled by the late Mr. Tom Eastwood, the author of the First Edition (1935) on which the Second and Third Editions (the latter with some additions by Dr. F. M. Trotter and Mr. W. Anderson) were closely based.

Since the First Edition was published much additional geological work has been done in the region, and many exploratory bores have been made, carrying forward our knowledge to the point where a new edition is deemed to be necessary and the work has therefore been almost wholly rewritten. There are still subjects for controversy, however, and on such topics this account presents the views that are thought to be most widely held at the present time. The authorities for them are not generally quoted in the text, but a select bibliography of the works on which they are based is given at the end of the book.

Our thanks are due to the Geography Department, University of Newcastle upon Tyne, for permission to reproduce the cover photograph and Plate XA and B, all taken by Mr. W. W. Anson; to Cambridge University for Plate VIA taken by Dr. J. K. St. Joseph; to Derek Crouch (Contractors) Limited for Plate XII; and to Dr. P. Beaumont for Plate XIB.

Thanks are also due to Mr. T. S. Tomlinson, North East Geological Outstation, National Coal Board, for information on the coal seams of northern England.

Chapters 2 and 4 of the work were prepared by Mr. Burgess, Chapter 3 by Dr. Warren and Mr. Burgess, the Carboniferous Limestone Series and Millstone Grit Series in Chapter 5 by Mr. Land and Mr. Mills respectively, Chapters 7 and 10 by Mr. Smith and Chapters 8, 9 and part of Chapter 6 by Mr. Eastwood. The remainder was prepared by Mr. Taylor who also edited the work. The map of the geology of the region, Plate XIII, was compiled by Mr. G. Richardson.

Institute of Geological Sciences,
Exhibition Road,
South Kensington,
London, S.W.7.
21st June, 1971

K. C. DUNHAM
Director

An EXHIBIT illustrating the geology and scenery of the region described in this handbook is set out in the Geological Museum, Institute of Geological Sciences, Exhibition Road, South Kensington, London, S.W.7.

Contents

Illustrations

Figures in Text

Plates

[1]Numbers preceded by A, MLD or L refer to photographs in the Geological Survey collections.

Plate   **Facing page**

1. Introduction

Geology and Landscape

The Northern England region includes the whole of Northumberland, Durham, Cumberland, the Furness district of Lancashire and the Isle of Man, most of Westmorland and a little of north Yorkshire (Plate II). The chief elements in its topography are the Cheviot Hills and Northumbrian Fells, the northern Pennines, the Cumbrian Mountains and Howgill Fells, the Solway Plain and Vale of Eden, and the lowlands of Northumberland and Durham. The main watershed of England, separating east- and west-flowing rivers, crosses the Scottish border in the Larriston Fells and the Tyne Gap near Gilsland, thereafter running close to the edge of the west Pennine escarpment until it crosses the Stainmore Gap.

The present distribution of high and low land has been strongly influenced by geological factors, some of which are of great geological antiquity and the result of long continued processes operating deep in the earth's crust. For instance, the Isle of Man, the Lake District and the northern Pennines lie along the axis of an ancient upfolded mountain chain, which was initiated at the close of Devonian times in the Caledonian orogeny. This elevated ground influenced sedimentation in the early Carboniferous because it formed a peninsula or island in the Carboniferous sea, and even when it was completely submerged and engulfed by later sediment, parts of it at least continued to act as 'positive' tectonic zones, tending to subside less rapidly than those parts of the crust immediately to the south and north. And to this day, the north-easterly trending axis of the Caledonian ridge, now broken by Hercynian and Tertiary earth movements into at least three segments—the Isle of Man, the Lake District and the northern Pennines—is still recognizable amidst the wider mass of high ground in the region.

Another geological control is exerted by variations in the hardness of the rocks and their resistance to erosion. Thus, the soft shales and relatively poorly cemented sandstones of the Permian and Trias generally crop out in valleys; and the unresistant shales of the Northumberland, Durham and Cumberland coalfields form relatively low ground relieved only by subdued escarpments formed by the thicker sandstones. By contrast, the thick Carboniferous limestones, the Fell Sandstones of Northumberland, the thicker Namurian and lowest Coal Measures gritstones and the Great Whin Sill, form the rugged and spectacular crags that are so much a feature of the region's sky-lines. The Permian Magnesian Limestone, flanked on one side by the sea and on the other by the relatively low-lying Coal Measures, forms a prominent westward-facing escarpment in Durham, and makes high vertical cliffs where it is attacked by wave action at the coast.

The most truly mountainous scenery is provided by the Ordovician rocks of the Lake District, hardened by metamorphism, which have maintained their relative elevation through many cycles of erosion during which all remnants of later rocks have been stripped away, leaving only the resistant core of the dome-like structure. In the tract included between lines drawn through Ullswater, along the eastern and southern sides of Derwentwater towards Ennerdale in the north, and through Coniston and Ambleside in the south, the rocks of the Borrowdale Volcanic Group give rise to the bold craggy heights which include Helvellyn, the Scafells and Langdales, and the famous screes of Wastwater. These form a marked contrast to the smoother, less rugged but still mountainous Skiddaw Slates country to the north, and to the milder type of scenery to the south, typical of the area occupied by Silurian rocks. The Isle of Man is like the Cumbrian Mountains in miniature, with a similar geological history. It has a central upstanding core of hardened Cambrian and Ordovician sediments, flanked to north and south by younger rocks. By contrast with the Lake District, however, the more rounded out-lines of the higher mountains reflect the absence of volcanic rocks com-parable with the Borrowdales.

Scenery in northern England has also been strongly influenced by the effect of the ice sheets which invaded the region, probably several times during the Pleistocene epoch. The whole region was overridden by ice, which has rounded to a greater or lesser degree the outlines of much of the upland country of the Pennines and Border. The ice and its meltwaters had a strongly erosive effect in the upland areas, especially in the Lake District which was a local centre of ice accumulation, and where in consequence the valleys underwent considerable modification. Wastwater, the bottom of which descends below sea level, is perhaps the best example of over-deepening. Hanging valleys are common and betray, by the amount of 'hang' or differ-ence in level with respect to the main valley, the amount of material that has been removed by ice. An indication of how great this amount was is given by the fact that a high proportion of the boulders in thick glacial tills that extend as far south as Shropshire in the west and Doncaster in the east are of Lake District rocks, such as Ordovician tuffs and Ennerdale and Shap granites.

Much detailed sculpturing of the land has been effected by the swift and highly erosive streams flowing beneath, and around the margins of the ice. These have created valleys that are in many cases much too large and deep for the streams that now occupy them. Rivers were diverted from their old courses when their valleys were blocked by ice, never to return to the former channels. Many of the deep drift-filled channels that have been proved in bores in Durham and Northumberland, some of them below sea level and once thought to be the buried river valleys of an older landscape, may have been cut by powerful streams flowing beneath the ice sheet, to be filled in later by glacial tills and the outwash products of a wasting ice-mass.

The topography of the valleys and coastal plains also owes much to the deposition of material from glaciers. If it were not for glacial deposits much of the Solway Plain, and parts of the West Cumberland and Furness coastal strip would be sea; likewise the Tees Estuary, which stripped of glacial deposits would extend inland almost to Darlington. Most of the low ground is covered by glacial drift of one sort or another, the characteristic surface expressions of which include rounded drumlins, hummocky moraines and kame belts, and featureless expanses of ground-moraine boulder clay.

Finally, geological processes in the period since the ice-caps receded from the region have left their mark. Great spreads of blanket peat have formed on the high wet Pennine and Border fells, and the thick deposits of lowland or basin peat which have formed on the more ill-drained parts of the lowland are best evinced in the Solway Plain north-west of Wigton. River terraces, composed of sediments varying from coarse boulder gravel to fine silt and mud, flank many of the rivers and by their various levels tell the history of the developing drainage patterns. Raised beaches and benches of marine warp, formed when isostatic recovery of the land after the removal of the ice-load lagged behind the rise in sea level due to the ice-melt, are a feature of both Irish Sea and North Sea coasts. The processes of freeze-and-thaw, scree formation, and landslip, have mantled many hillsides with unconsolidated material which smooths out the landscape, concealing features in the more solid rocks beneath.

Geological History

The geological history of the region, so far as it can be based upon the tangible evidence of the rocks, begins towards the close of the Cambrian Period and the outline of these events is shown in Table 1.

During this vast period of time—about 500 million years—the region has been the scene of a wide variety of environments, ranging from arid and almost lifeless desert to warm sub-tropical seas teeming with marine creatures, some of which have left their remains and impressions in the rocks to provide an evolutionary record for palaeontologists. The climatic changes were, in large part, local manifestations of processes that affected the whole earth, like the movements of the polar axes which placed our region nearer or farther from the equator, the relative movement of the main land masses under the influence of continental drift, variations in the composition of the atmosphere, and world-wide temperature variations such as those that occurred during ice ages. The changes in the distribution of land and sea, which have resulted in such a varied sequence of layered rocks of enormous total thickness, are the result of changing regional crustal stresses which have been responsible for a repeated cyclic sequence of subsidence and marine deposition, followed by folding, uplift and erosion. This sequence has been repeated on a major scale at least four and possibly five times since the Cambrian. Thus, the general structural pattern in the rocks is one of increasing complexity and deformation with increasing age, the older rocks being more hardened and altered by metamorphic and diagenetic processes.

At various times from the Ordovician onwards igneous rocks have been intruded into the sedimentary strata in the form of plutons, dykes and sills, and at times the intrusions broke through to the surface as volcanoes, resulting in the accumulation of thick layers of ash or lava.

The results of the above processes can be examined in and around the Lake District, as Fig. 1 shows graphically.

The recorded geological history of the Northern England region starts as the mud and silts of the Manx Group (Fig. 3) were being laid down in a wide subsiding geosyncline lying between a great land platform to the north-west, occupying the relative position of the present north Atlantic, and land masses to the south. This basin continued to subside and to receive the

TABLE 1

Periods	Deposits	Conditions and Events
RECENT	Blown sand, peat, alluvium, raised beaches, marine warp, submerged forests	Relative changes in sea level; amelioration of climate
PLEISTOCENE (·01 – 2 m.y.)	Boulder clay, sand, gravel, laminated clay	Glacial: invasion of the region by ice-sheets, and their recession; probably on several occasions
	Unconformity	Great erosion
TERTIARY (2 – 65 m.y.)	Igneous dykes (no sediments present)	Uplift of the whole region, folding, much faulting
CRETACEOUS (65 – 136 m.y.)	No sediments present	
JURASSIC (136 – 195 m.y.)	Shale and limestone (top of succession absent)	Marine
TRIASSIC (195 – 225 m.y.)	Red mudstone and siltstone with rock salt, and sandstone	Semi-arid
PERMIAN (225 – 280 m.y.)	Limestone and dolomite, sandstone, anhydrite, gypsum, rock salt, sandstone and breccia	Sub-tropical seas margined by semi-arid desert
	Deposition of lead and zinc ores and associated minerals in northern Pennines	
	Unconformity	Folding, great erosion
	Intrusion of Whin Sill and igneous dykes	
CARBONIFEROUS (280 – 345 m.y.)	Shale, sandstone, coal (chiefly in upper part), limestone (chiefly in lower part), conglomerate, basalt	Deltaic and marine; some active volcanoes
OLD RED SANDSTONE (345 – 395 m.y.)	Granite and other plutonic igneous rocks; lava and agglomerate; conglomerate	Volcanoes (in Cheviots)
	Unconformity	Severe folding, great erosion
SILURIAN (395 – 440 m.y.)	Mudstones and sandstones	Marine; final infilling of sedimentary basin
ORDOVICIAN (440 – 500 m.y.)	Shales (ashy in part) calcareous shale and limestone, conglomerate, tuff, lava, agglomerate	Geosynclinal subsidence, marine, some volcanoes
	Unconformity	Folding and erosion
	A thick sequence of lava and tuff	Volcanoes (Lake District)
	?Unconformity	?Folding and erosion
	Mudstone, siltstone, sandstone	
?CAMBRIAN	Mudstone, siltstone (lowest Manx Group)	Marine, geosynclinal

m.y. = million years before present

sediments of the Ordovician Skiddaw Group until the earlier part of Llanvirn times (p. 12), when the eruption of tuffs and lavas in the eastern part of the Lake District, and in the Cross Fell Inlier, followed and possibly accompanied by some uplift and erosion, foreshadowed a much more violent and extended series of volcanic eruptions which were centred mainly on the Lake District, their products accumulating as the Borrowdale Volcanic Group. The volcanoes rapidly built up their cones well above sea level, so that most of the debris, which included ashes, lavas, and the products of *nuées ardentes*, fell

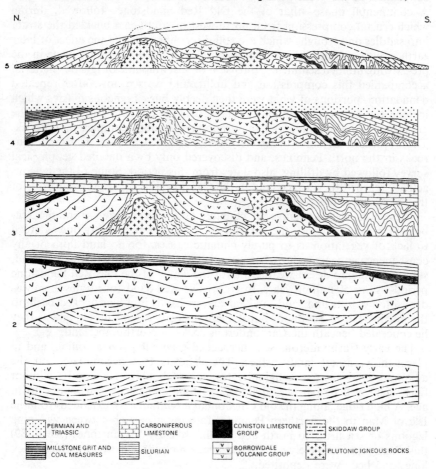

PERMIAN AND TRIASSIC

CARBONIFEROUS LIMESTONE

CONISTON LIMESTONE GROUP

SKIDDAW GROUP

MILLSTONE GRIT AND COAL MEASURES

SILURIAN

BORROWDALE VOLCANIC GROUP

PLUTONIC IGNEOUS ROCKS

FIG. 1. *Diagrammatic sections to illustrate the building of the Lake District*

1. Deposition of Skiddaw Group; folding and erosion; deposition of Borrowdale Volcanic Group.
2. Folding and erosion; deposition of Coniston Limestone Group and Silurian rocks.
3. Severe folding and great erosion; intrusion of plutonic igneous rocks; deposition of Carboniferous rocks.
4. Gentle folding and considerable erosion; deposition of Permian and Triassic rocks.
5. Gentle uplift, producing an elongated dome and resulting in radial drainage; erosion to present form.

on land and suffered considerable erosion; but subsidence was then resumed, the sea once more drowned the region, and the calcareous marine sediments of the Coniston Limestone Group were laid down unconformably upon both Borrowdale Volcanics and Skiddaw Slates, though with a brief recrudescence of volcanic activity during which the Stockdale or Yarlside Rhyolite was formed.

Silurian muds, silts and sands followed without a break, but by the end of the period the geosyncline had narrowed, and earth movements which culminated in the great period of folding termed 'Caledonian' had begun. A continental phase—that of the Old Red Sandstone—followed, during which crustal compression from north-west or south-east buckled the strata into saddles and troughs which are still discernible, despite the effect of later earth movements, and this compression also induced a slaty cleavage in the pre-existing lithified sediments. Intrusion of vast amounts of igneous material accompanied this compression and uplift, and we see now, after repeated denudation, portions of what were then deeply buried igneous masses which include the granites of Shap, Skiddaw, Eskdale and the Isle of Man, the granophyre of Ennerdale and the granophyre-gabbro complex of Carrock. The Weardale Granite, completely concealed beneath the Carboniferous rocks in the north Pennines, and discovered only by a detailed geophysical survey followed by drilling, also dates from this period. Some of the igneous activity reached the surface, and in the Cheviot Hills we see the remnants of the volcanoes of this period.

There was also a change in climate; land newly made from the sea-bottom became a desert, the barrenness of which was possibly due as much to lack of vegetation as to purely climatic causes, for no land flora of any consequence appears to have been in existence. The region remained as land, subject to intense desert weathering for the whole of the Old Red Sandstone period—some 50 million years. In Upper Old Red Sandstone times there was deposition of silts and sands and calcareous bands in a land-locked basin to the north-west, in Scotland, but if these deposits extend into England they lie concealed beneath the Carboniferous rocks of Northumberland.

The early Carboniferous sea encroached gradually over a rolling, and in places hilly, desert land surface. The deposits that accumulated on the adjoining land—coarse conglomerates, screes, boulder beds and piedmont fans—may well be partly Old Red Sandstone and partly Carboniferous in age. A high ridge, along the old Caledonian mountain axis, extended from the Isle of Man to the northern Pennines forming a barrier between widening lagoons or sea firths to north and south. By early Viséan times (see p. 39) this ridge was completely submerged beneath clear shelf seas in which thick limestone beds were deposited.

Volcanic eruptions during the early part of the Carboniferous period were responsible for, first, the basaltic lavas of the Border region, then those of Cockermouth, and lastly those of the Isle of Man. These indicate a westward shift of volcanic outbreaks during Lower Carboniferous times.

There remained a tendency for the old-established Caledonian elevations to persist as 'positive' areas or 'blocks' where subsidence was less than in the intervening parts of the crust, particularly where large volumes of in-truded granite imparted isostatic buoyancy, as in the Lake District, the north Pennines and the Southern Uplands of Scotland. The result was that sequences of strata laid down in the intervening basins were thicker than their time-equivalents over the 'blocks'. This tendency for differential subsidence persisted throughout the Namurian, but by Westphalian times it had almost ceased. The supply of detrital material from land to the north was more than sufficient through Upper Carboniferous times to keep pace with the subsidence of the region, and a wide deltaic area was established, on the swamp surface of which land vegetation flourished. The repeated drowning, and rebuilding to sea level of this surface by deposition gave rise first to the rhythmic Yoredale sequences and later to the characteristic Namurian and Coal Measures cyclothems.

The deposition of red Upper Coal Measures in Cumberland foreshadowed the approaching aridity of Permo-Triassic times. The folding which brought the Carboniferous Period to a close continued into early Permian times. It was responsible for the initiation of the Pennines as a north–south ridge with major faults to west and north; for some upwarping in the Lake District, and for the isolation of the coalfields, the remnants of which are preserved in synclines initiated at this time.

Probably towards the end of the Carboniferous the whole region was again uplifted to bring in a second continental phase, which like the earlier one also had its igneous activity. The intrusion of the Great Whin Sill, in places in more than one leaf, into Carboniferous strata beneath some 5000 sq km of northern England, together with the emplacement of the associated dykes, represents the major igneous event of this time. The uplifted Carboniferous rocks were deeply eroded and reddened from the surface by desert weathering, and among the earliest Permian deposits were breccias or 'brockrams' mantling the rising ground. Wide areas of the land were covered by sand dunes with their characteristic 'millet seed' sand grains, rounded by the action of the wind. Transgression by the *Bakevellia* and Zechstein seas, in conditions of high temperature and low humidity, then brought about the formation of the Magnesian Limestone. At first, the sea supported a dwarf shelly fauna, but with increasing salinity, which led to the accumulation of thick anhydrite and gypsum beds on both sides of the Pennines and of rock salt in Durham, the basin became almost devoid of life apart from mat-forming algae.

Throughout the Triassic Period deposition of red sands and red and green mottled muds continued in shallow 'dead' seas surrounded by deserts. As in the Permian, evaporation was high enough to cause the precipitation of the calcium sulphates anhydrite and gypsum, and rock salt. There is insufficient evidence to decide whether or not a significant barrier existed in Triassic times along the Pennine line between deposits in the east and west, but it is possible that the uppermost Triassic deposits, at any rate, may have once extended over much of the region including the present Pennines. The Keuper Marl of the east is remarkably similar to that of Furness, and to the contemporaneous Stanwix Shales of the Solway Plain. Some geologists believe that these red muds are in large part wind-blown material transported in dust storms from neighbouring land into standing water.

The desert régime in northern England was ended by widespread submergence beneath the Rhaetic sea. How much of our region was covered and how much remained as land is a matter for conjecture, for the only

Mesozoic sedimentary rocks younger than the upper Triassic are the Lower
Lias limestone and shale, preserved from denudation in a shallow syncline near
Carlisle, and the presence of Rhaetic Beds beneath them is inferred simply
from the fact that Rhaetic Beds occur at all other places in the British Isles
where Keuper Marls are overlain by the Lower Lias. The direct record of
the 'solid' rocks finishes with this Lower Lias. Some of the later formations
were probably deposited, but were removed by denudation following the
great earth movements in Tertiary times.

These movements gave rise to much of the faulting, and are mainly
responsible for the arrangement of the rocks as we now see them. To them we
really owe the Lake District, for they raised the rocks there into a great dome
from which thousands of metres of newer beds were weathered away,
leaving Carboniferous, Permian and Triassic outcrops disposed around a
core of already folded rocks. The consequent radial drainage initiated on
this dome has persisted, though the valleys, and the low ground generally,
have been modified by the action of ice. Farther east the Pennine escarpment
of Carboniferous Limestone rises from the low-lying Permo-Triassic rocks.
The latter formations are the younger, so the Pennine Fault between them
must be younger still. There is evidence of faulting along this line at much
earlier periods, but the movement was probably completed during Tertiary
times and was responsible, together with some denudation, for the Vale of
Eden lying between the raised block of limestone country to the east and the
Lake District to the west.

The eastward tilting of the north Pennines, with movement on the east to
west Stublick–Ninety Fathom fault system, was another consequence of
Tertiary earth movements. The tendency for certain parts of the pre-Carboniferous basement, particularly where large masses of granite had been intruded
in Caledonian times, to act as 'positive' or buoyant areas which subsided less
rapidly than other parts of the region was again manifest in Tertiary times
(Fig. 2). The latest, as well as the earlier, movements along some of the
major faults, including those along the Pennine, Stublick–Ninety Fathom,
and Lunedale–Butterknowle lines, may be regarded as affording relief to
these differential stresses.

These major crustal movements, like the earlier ones, were accompanied
by igneous activity. Elsewhere—in the west of Scotland and the Inner
Hebrides, and in Northern Ireland—Tertiary igneous activity was on a large
scale, and included plutonic, hypabyssal and volcanic phases. In northern
England the chief manifestation of this is in a suite of east-south-easterly
heading tholeiitic dykes, the most northerly of which is the Acklington Dyke
and the most southerly the Armathwaite–Cleveland Dyke. Both of these
trend towards the Tertiary volcanic province of Scotland. Some of the
olivine-dolerite dykes of the Isle of Man also belong to this period. Apart
from the evidence of earth movements, whose Tertiary age is inferential, and
the igneous dykes, we have no information from the region about events in
this era.

The ensuing Quaternary era which embraces the Pleistocene and Holocene
(or Recent) periods was unlike any previous time for which we have evidence
within the region. Whether because of world-wide climatic change or the
shifting of the earth's rotational axis—more probably the former—the edge

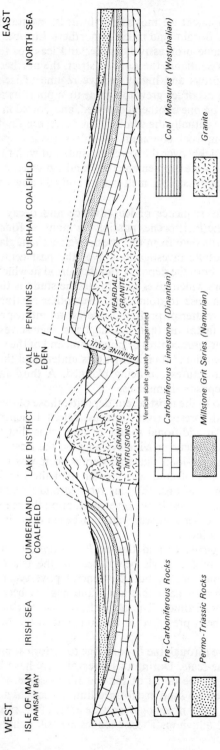

Fig. 2. *Diagrammatic section from the Isle of Man to the North Sea*

Note: 1. The formation of the synclines in which Coal Measures are preserved by pre-Permian folding, and their tilting by post-Permian movements.
2. The large granite plutons that have imparted isostatic buoyancy to the Lake District and northern Pennines.
3. That the Pennine Fault, though shown here as largely post-Permian, was initiated at an earlier stage.

of the polar zone of permanent ice moved southwards, eventually covering all but a southern strip of Britain, and much of northern Europe. During the advance, the climate became progressively colder and ice-caps formed over regions of high ground in Scotland, the Lake District, the northern Pennines and the Cheviots. A vigorous and highly erosive régime of ice-flow set in as the accumulated thickness of ice grew too great to support its own weight. The ice streams impinged on one another, coalesced, and flowed in a generally southerly direction until eventually the whole of the Northern England region was overridden. Ground-up rock containing boulders was transported many kilometres and smeared thickly over the lower ground, and with the waning of the ice sheets these deposits were supplemented by sands, gravels and laminated clays, washed out of the ice by copious melt-waters and laid down around and beneath the ice-margins.

It seems likely that the sequence of ice advance and decay took place more than once during the Pleistocene, possibly as many as four times, but most of the known deposits date from the latest, Weichselian, glaciation. So complex are they that despite investigations over the past century we still have much to learn about both the deposits and the period in which they were formed. One of the factors which gives impetus to the study is that this was the period in which Man rose to dominance over other creatures, and his early environments are a matter of profound interest to anthropologists and pre-historians. Another factor is that the deposits themselves form the foundations of most of our large centres of population and of the motor roads linking them, and it is essential that civil engineers understand their physical properties when heavy structures are being planned. A third factor is the economic value of the deposits themselves.

These factors apply also to the most recent deposits, those of the Holocene, which include submerged forests, river terraces, raised beaches and platforms of marine warp. An additional source of interest is that the varying levels at which the deposits occur provide evidence on the isostatic recovery of the land after the huge ice-load that had depressed it was removed. At times when the isostatic rise of land was greater than the rise in sea level due to the melting of the ice-caps, the seas receded and forests flourished far out from the present coastlines. In the reverse situation the sea encroached upon the land, and marine benches were cut and beach deposits laid down high above the present sea level.

The Irish Elk *Cervus giganteus* and the ox *Bos longifrons* flourished in post-glacial times; some of the marls below peat in the Isle of Man are famous for remains of the former, but the general poverty of the Manx post-glacial fauna and flora suggests that the island has not been connected to the mainland since glacial times. The horn-sheaths of *Bos taurus taurus* and *B. primigenius* have been preserved in blanket peat in the Moor House area of Westmorland.

From the Neolithic Age through the Bronze Age to historical times is but a step in the geological time-scale. During this interval there have been minor changes of sea level and in the detail of the coastlines, some of the ponds and lakes have silted up, blown sand has formed dunes, peat has accumulated in ill-drained basins and as blanket bog over wide areas of the rainy uplands, and rivers have deposited alluvium. These changes are still going on.

2. Ordovician System

The Ordovician rocks of northern England were deposited within the confines of the Caledonian geosyncline, and both a deep-water open-sea facies with greywackes and graptolitic shales, and a shallow-water near-shore facies, with shelly limestones and shales are represented. A great thickness of contemporaneous volcanic rocks is also present. The Ordovician rocks crop out in the Isle of Man, the Lake District, and in inliers at Cross Fell and Teesdale (see Plate XIII), and they are believed to have been penetrated beneath Carboniferous strata in borings at Crook and Allenheads in County Durham. A tabular summary of the stratigraphy is given in Fig. 3.

Manx Group

The Ordovician rocks of the Isle of Man (Fig. 12) are the oldest in the region. They comprise a thick sequence (7500 m) of alternating greywackes, siltstones and mudstones, with several beds of slump breccia. Microfossils obtained from the Lonan Flags near the base of the Group suggest either a late Tremadoc or early Arenig age. The only graptolites so far discovered are dendroid forms (*Dictyonema*) from the Cronkshamerk Slates, near the top of the sequence.

Skiddaw Group

In the main Lake District outcrop (Fig. 4), these rocks closely resemble those of the Manx Group in lithology and facies and are probably the result of continued deposition within the same sedimentary basin. They differ in the not uncommon occurrence of graptolites of Arenig and Llanvirn age, indicative of the following ascending sequence of zones: *?Tetragraptus approximatus, Didymograptus extensus* (subzones of *D. deflexus, D. nitidus* and *Isograptus gibberulus*), *D. hirundo, D. bifidus, D. murchisoni.* Microfossils such as acritarchs and chitinozoa have also been of use in correlation.

The general strike of the beds is Caledonoid, and across this north-east–south-west regional trend there is an alternation of flaggy and muddy formations, several of which have been given local names such as Blake Fell Mudstones, Loweswater Flags, Kirkstile Slates and Mosser Slates. Whether or not these represent discrete horizons is still a subject of controversy. The beds have been subjected to several phases of compression and are in places so highly cleaved and contorted that the relationship between neighbouring formations is open to conflicting interpretations. The two extreme views of the regional structure are:

(a) that the entire sequence is about 1800 m thick, comprising unfossiliferous shales at the base, overlain by grits and flags with graptolites (the Loweswater Flags) which become finer grained upwards till they grade into siltstones and mudstones, also fossiliferous (the Mosser–Kirkstile Slates). In this view the successive outcrops of flags-and-shale sequences are the result of repetition by tight folding of the same few horizons (Jackson 1961–62; Eastwood and others 1968).

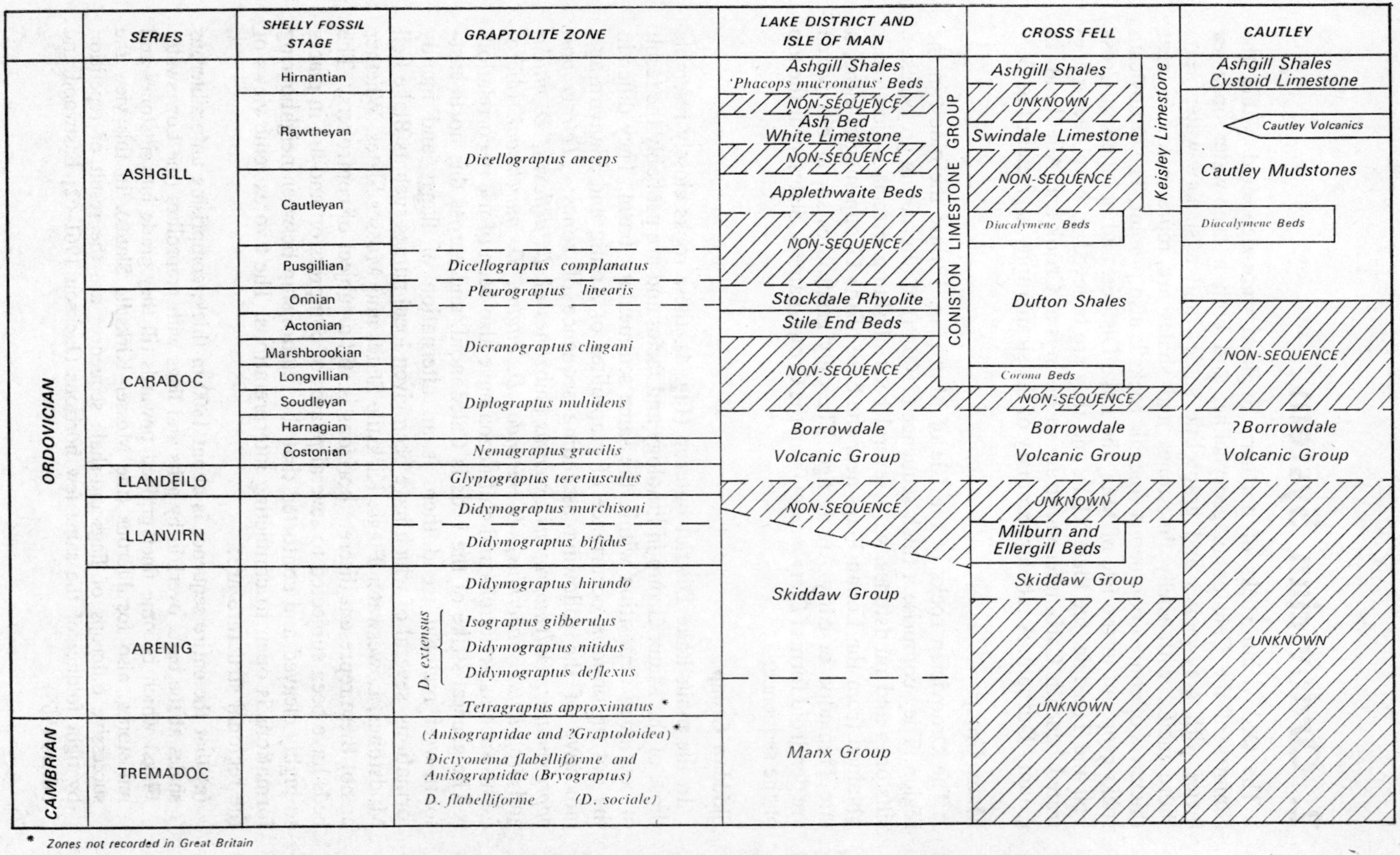

	SERIES	SHELLY FOSSIL STAGE	GRAPTOLITE ZONE	LAKE DISTRICT AND ISLE OF MAN	CROSS FELL	CAUTLEY
ORDOVICIAN	ASHGILL	Hirnantian	*Dicellograptus anceps*	Ashgill Shales / 'Phacops mucronatus' Beds	Ashgill Shales / Keisley Limestone	Ashgill Shales / Cystoid Limestone
		Rawtheyan		NON-SEQUENCE / Ash Bed / White Limestone	UNKNOWN / Swindale Limestone	Cautley Volcanics / Cautley Mudstones
		Cautleyan		NON-SEQUENCE	NON-SEQUENCE	
		Pusgillian	*Dicellograptus complanatus*	Applethwaite Beds	*Dicellograptus* Beds	*Dicellograptus* Beds
	CARADOC	Onnian	*Pleurograptus linearis*	CONISTON LIMESTONE GROUP	Dufton Shales	
		Actonian	*Dicranograptus clingani*			
		Marshbrookian	*Diplograptus multidens*	NON-SEQUENCE	*Corona* Beds	NON-SEQUENCE
		Longvillian		Stockdale Rhyolite	NON-SEQUENCE	
		Soudleyan		Stile End Beds		
		Harnagian	*Nemagraptus gracilis*	NON-SEQUENCE	Borrowdale Volcanic Group	?Borrowdale Volcanic Group
		Costonian	*Glyptograptus teretiusculus*	Borrowdale Volcanic Group		
	LLANDEILO		*Didymograptus murchisoni*		UNKNOWN / Milburn and Ellergill Beds	
	LLANVIRN		*Didymograptus bifidus*	NON-SEQUENCE		UNKNOWN
			Didymograptus hirundo	Skiddaw Group	Skiddaw Group	
	ARENIG		*Isograptus gibberulus*			
			Didymograptus nitidus		UNKNOWN	
			Didymograptus deflexus ⎱ *D. extensus*			
			Tetragraptus approximatus ⎰	Manx Group		
			(Anisograptidae and ?Graptoloidea) *			
CAMBRIAN	TREMADOC		*Dictyonema flabelliforme* and Anisograptidae (*Bryograptus*)			
			D. flabelliforme (*D. sociale*)			

* Zones not recorded in Great Britain

Fig. 3. *Table summarizing Ordovician stratigraphy in northern England*

(b) that the sequence comprises an alternation of coarse- and fine-grained formations totalling 9000 m in thickness, with the beds disposed in broad anticlines and synclines—the small folds seen in outcrops being minor structures on the limbs of the major folds (Simpson 1967).

There is as yet insufficient evidence from fossils to show which of the above structural interpretations is the more likely.

Rocks of probable Arenig age crop out in inliers at Black Coombe and Greenscoe. They are also seen in the Cross Fell Inlier, where they form the pikes of Murton and Brownber.

Rocks of Llanvirn age crop out in the northern part of the Cross Fell Inlier. They consist either of graptolitic mudstones ('Ellergill Beds') or of an alternation of similar beds with basic, water-laid tuffs and some lavas ('Milburn Beds'), the whole sequence totalling many hundreds of metres, all lying within the *Didymograptus bifidus* Zone. Mudstones and volcanic tuffs of this age (the "Mottled Tuffs" of J. F. N. Green), are also present in the eastern part of the Lake District, in the Bampton and Ullswater inliers, and mudstones of the overlying *D. murchisoni* Zone have recently been recorded in this area. In the west, these beds are absent, and volcanic rocks apparently rest directly on slates of Arenig age.

Borrowdale Volcanic Group

In the Lake District the Skiddaw Group is succeeded by a great thickness of mainly subaerial volcanic rocks—lavas, tuffs and agglomerates, totalling in excess of four thousand metres—with some igneous intrusions. The junction between the two groups in some places is at a fault; in others it is an unconformity with a basal conglomerate (Bampton Conglomerate; Latterbarrow Sandstone). Some geologists consider that deposition of the Borrowdale Volcanic Group followed that of the Skiddaw Slates without a break. However, evidence is now accumulating that the two groups are separated by a period of folding, uplift and erosion; and moreover that the tuffs and lavas of the Skiddaw Group are not the immediate precursors of the Borrowdale Volcanics, but the products of a quite distinct, much earlier volcanic episode.

No fossils have been found in the Borrowdale Volcanic Group. As these rocks are succeeded in the Cross Fell Inlier by strata of Caradoc (Longvillian) age, their general equivalence to the Llandeilo Series of Wales has been inferred; and they may in part be of lower Caradoc age.

The petrology and stratigraphy of the rocks of the Borrowdale Volcanic Group have been summarized by Dr. G. H. Mitchell and the following account is derived mainly from his work.

The sequence consists of a thick pile of pyroclastic rocks varying from extremely fine-grained tuffs to agglomerates, interbedded with which are numerous flows of lava ranging in composition from basalts to rhyolites, with andesites as the dominant variety. As might be expected in such an accumulation the various flows and tuff bands are often impersistent and this feature, combined with the absence of ordinary sedimentary material and the almost complete lack of fossil remains, makes the tracing of individual strata difficult, particularly in areas where exposures are infrequent or the ground is mantled with drift. Reliance on lithology as the principal means of comparison among the Borrowdale rocks introduces a further difficulty, for it becomes increasingly apparent that closely similar lavas and tuffs were repeatedly erupted at different times during the history of the volcanic pile and consequently appear at widely different horizons.

A striking character of the lavas is the flow-brecciation which occurs in both rhyolites and andesites. This may affect both the top and bottom of the flows as well as the front and is due to the welding of broken congealed crusts by still molten lava. Vesicular and massive lavas also occur, the former structure being particularly common among the more acid andesites while the latter is frequently a feature of augite-andesites. Vesicles are usually filled with chlorite, calcite, epidote, or quartz. Porphyritic types of lava are well represented though phenocrysts of feldspar or ferromagnesian minerals are usually small, exceptions being in the Eycott lavas. Many flows are distinctly fine grained or aphanitic. Despite devitrification, glassy varieties still show conchoidal fracture and perlitic cracks. A conspicuous platy jointing along the flow lines characterizes both acid and basic lavas in places and sometimes can only be distinguished with difficulty from the bedding of tuffs. Many lavas show flow-banding marked by colour bands, lines of vesicles, or changes in grain. Rhyolites are nodular in places. In colour the andesites are usually bluish green, green, grey, or even reddish purple, the more basic rocks being as a rule of a darker shade; rhyolites on the other hand are commonly pale grey, purple, greenish brown, or even yellow in appearance. The upper and lower surfaces of flows are often irregular and this is most conspicuous when they are in contact with tuffs. Instances are known of lavas penetrating tuffs and later deposits of tuffs are frequently much intermingled with the underlying flow-breccia.

The remarkable feature of the great thickness of Borrowdale Volcanic rocks is that, apart from the presence of lithic fragments in some of the basal beds, non-volcanic sedimentary material is absent throughout the group. It is not surprising, therefore, that fossils are almost unknown, the only remains being certain tracks, perhaps made by arthropods, described from the Lickle Valley.

A number of distinct types of tuff are common. The most readily recognized are the well-bedded, fine-grained lithic tuffs, green or grey in colour, which when cleaved provide the famous Westmorland or Cumberland green slates—though a large proportion of these is now quarried at Broughton Moor in Lancashire. The finer tuffs vary in grade from extremely fine halleflinta-like beds, particularly common in the neighbourhood of Sty Head, Esk Hause and Great Langdale, to sandy varieties often delicately bedded, a feature which shows very well on wet surfaces, particularly if they are cleaved. Many of them show ripple-marking and sometimes even current-bedding and slump-structures. A high percentage of carbonate is a character of some tuffs. Many fine-grained tuffs contain scattered lapilli or bombs.

Andesitic lithic tuffs with a considerable proportion of crystals are most common, but fine-grained rhyolitic tuffs are also known. In the latter, bedding is often not so well developed and they are difficult to distinguish from fine-grained lavas. These rhyolitic tuffs are generally coloured yellow or grey, in contrast to the green or bluish green, andesite varieties. Tuffs without obvious bedding are by no means unusual.

(A 2788)

A. The smooth outlines of Skiddaw Slate scenery: High Hows and Owsen Fell, Lamplugh. Kettle-moraine in foreground

Plate III

B. Rugged Borrowdale Volcanics scenery: Copeland Forest, seen from Wastwater

(MLD 6154 F)

A. Folding in metamorphosed Skiddaw Slates: River Caldew above Swineside

(A 6673)

Plate IV

B. Columnar jointing in andesite lavas: near Gosforth

(A 6716)

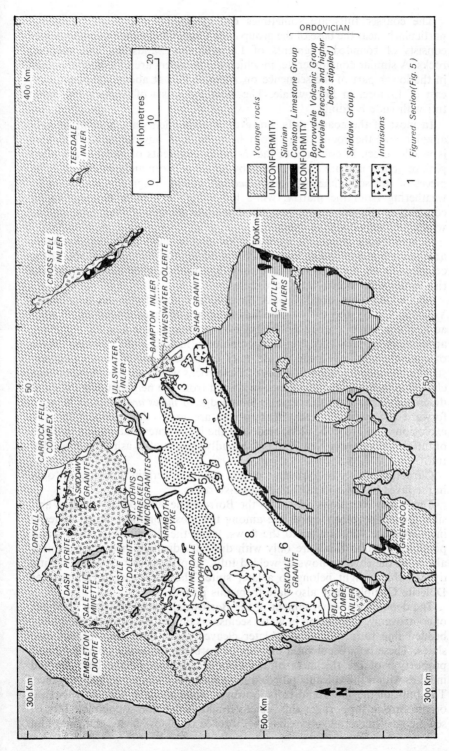

FIG. 4. *Distribution of Lower Palaeozoic rocks and major igneous intrusions in the Lake District, Cross Fell Inlier and Teesdale Inlier*

The coarser tuffs are sometimes distinguished by rounded fragments, particularly near the base of the group as at Capel Crag where a conglomerate consists of rounded fragments of Latterbarrow Sandstone and volcanic rocks. A similar conglomerate, in which the boulders are of andesite, occurs in the lower part of the sequence on Ashness Fell at about 460 m altitude, east of Lodore, in Borrowdale. The matrix consists of fine-grained, lithic tuff with rude bedding.

In most of the coarse tuffs, however, the fragments are angular or sub-angular and the rocks range from lithic tuffs to agglomerates. The base is usually green in colour, but the included fragments of rhyolite are often pink or yellow and are accompanied by others of green andesite as well as by pieces of tuff sometimes obviously bedded. Harder fragments stand out by weathering from the matrix; while the softer ones form hollows. Other conspicuous fragments are ragged, shard-like masses of deep blue colour which are probably 'lava-drops'.

Concretionary structures are well developed in many tuffs; they probably arose as a result of hydrothermal activity during the volcanic episode. Other concretions are the product of 'volcanic hailstones'; when cleaved they give rise to the well-known 'bird's-eye slates'.

For a long time streaky rocks have been known from the Borrowdale Volcanic Group and their origin has been a subject of discussion. Rather acid in composition they generally show a streaky banding having every appearance of flow-lines, yet they commonly include tuff fragments; these of course might have been accidentally introduced into a flow. Columnar structure is also a feature of some of these rocks which have sometimes been classed as tuffs and at other times separated from the neighbouring tuffs and described as rhyolites. Increase in knowledge of *nuées ardentes* has brought forward the suggestion that these rocks are really the products of such explosions and are to be compared with the ignimbrites and similar varieties of tuff described from New Zealand, broadly spoken of as ash-flow or welded tuff.

Garnet, probably of pyrogenic origin, occurs in certain districts, not only in the lavas but also in tuffs and intrusive rocks.

The minor intrusions related to the Borrowdale suite occur as dykes and sills. Quartz-porphyry dykes are among the most common. Small andesite intrusions are very similar to the flows, while the rhyolite sills, dykes, and possible vent-intrusions are only with difficulty distinguished petrographically from the rhyolite flows or welded tuffs. In places there are a few spilitic dykes, which might belong to a later suite. The much larger Haweswater Dolerite Complex may also date from this period.

The depositional environment of the Borrowdale rocks has been a subject of controversy. Deposition in water seems certain in the case of the delicately bedded fine tuffs which, when later cleaved, have been transformed into slates. These tuffs occur on a number of horizons but make up only a small proportion of the thickness of the Group. Some may have been formed in lakes within the volcanic pile or during periods of temporary submergence. The bulk of the flows and tuffs, however, seem to have been of subaerial origin. Strong support for this view is found in the almost complete lack of sedimentary material or fossils, the absence of pillow lavas, the rude and

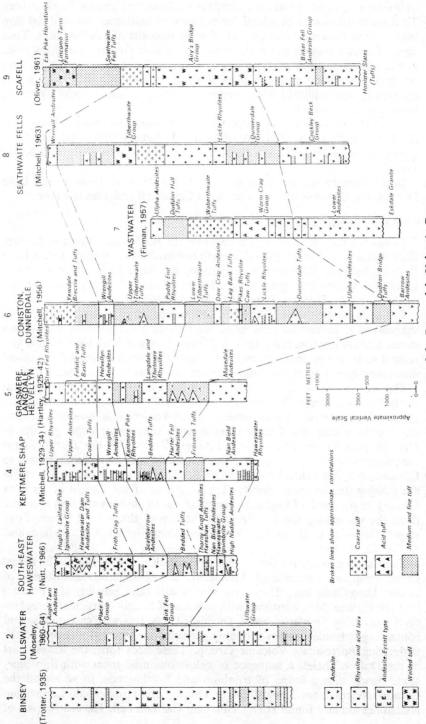

FIG. 5. *Comparative sections of the Borrowdale Volcanic Group*

unbedded character of many tuffs and the washouts and channelling in others. To these features may be added the presence of reddened and oxidized flow breccias and the recent recognition of many deposits of ash-flow tuffs. Then again the great erosion of the Borrowdale Volcanic Group which took place before the deposition of the Coniston Limestone Group seems to have occupied but a short period of geological time. The speed with which it was apparently accomplished also suggests that the volcanic rocks originally formed a subaerial pile.

From the very nature of these rocks, a uniform succession throughout the Lake District is scarcely to be expected. Representative sections from several recently revised areas are shown in Fig. 5, and their localities in Fig. 4.

In addition to the outcrops around the Skiddaw anticline in the Lake District, the Group is also present in the Cross Fell and Teesdale inliers.

Coniston Limestone Group

Following the extrusion of the rocks of the Borrowdale Volcanic Group, there was a period of considerable earth movement and erosion. In the Lake District, between Kentmere and Furness, the basal beds of the Coniston Limestone Group transgress some 2500 m of volcanic rocks coming eventually to rest on Skiddaw Slates at Greenscoe (Fig. 4).

The beds comprising the Coniston Limestone Group are mainly a series of shallow-water calcareous sediments, but include conglomerates, ashy sandstones and contemporaneous tuffs and lavas. Strata of both the Caradoc and Ashgill series are present, although, due to the presence of intra-formational unconformities, a complete sequence does not occur in most areas.

The local zonation of the Caradoc and Ashgill series (Fig. 3) is based on shelly fossils, as graptolites are uncommon; and the exact correlation with the standard graptolite zonal sequence is not known. It is, however, general practice to equate the Caradoc Series with the *Nemagraptus gracilis* to *Pleurograptus linearis* zones and the Ashgill Series with the *Dicellograptus complanatus* and *D. anceps* zones. The Caradocian stages from Costonian to Onnian were first defined in the Welsh Borderland, where succeeding strata are missing due to unconformity. The type area of the Ashgill Series is at Cautley, in the north of England, where there is an almost unbroken sequence from the underlying Onnian Stage to the top of the Ashgill Series. Four stages, Pusgillian to Hirnantian, have been defined by Dr. J. K. Ingham and Dr. A. D. Wright.

The Caradoc Series is thickest (325 m) in the Cross Fell Inlier. The oldest strata, cropping out around Melmerby, are fossiliferous mudstones of Lower Longvillian age. They pass southwards (and possibly shorewards) into the *Corona* Beds, siltstones with a fauna of mainly inarticulate brachiopods (including *Trematis corona* Davidson), bivalves and gastropods, and containing volcanic detritus probably derived from nearby outcrops of the underlying Borrowdale Volcanic Group. These beds form the lowest part of the Dufton Shales, a sequence of calcareous mudstones with thin limestone bands with a fauna of trilobites and brachiopods, in which all the Caradoc stages from Longvillian to Onnian are represented. Longvillian mudstones are also found at Dry Gill, near Carrock Fell. The central part of

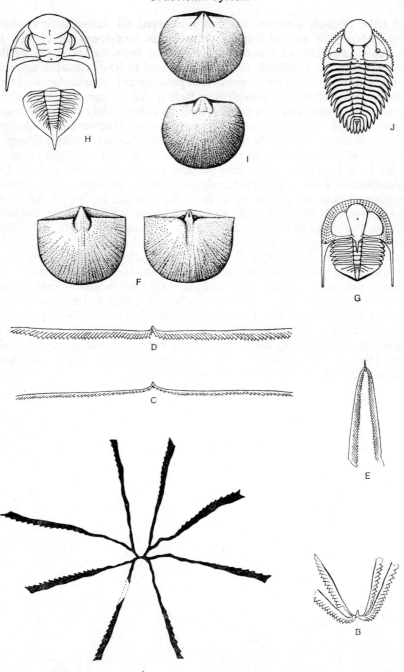

FIG. 6. *Ordovician Fossils*

(All drawings except J are natural size. Graptolites chiefly after Elles and Wood.)

Skiddaw Slates: Arenig Series—**A,** *Dichograptus octobrachiatus* (Hall); **B,** *Tetragraptus serra* (Brongniart); **C,** *Didymograptus extensus* (Hall); **D,** *Didymograptus hirundo* Salter; Llanvirn Series—**E,** *Didymograptus bifidus* (Hall).

Caradoc Series—**F,** *Dolerorthis duftonensis* (Reed); **G,** *Broeggerolithus nicholsoni* (Reed).

Ashgill Series—**H,** *Dalmanitina mucronata* (Brongniart); **I,** *Hirnantia sagittifera* (McCoy); **J,** *Staurocephalus clavifrons* Angelin, ×2.

the Lake District, however, was not submerged till Actonian times (Stile End Beds), while around Cautley only Onnian mudstones are present. In the main Lake District outcrop, the Stile End Beds are succeeded by the contemporaneous Stockdale Rhyolite, presumed to be of Caradoc age.

In the Cautley area, Onnian mudstones pass upwards in unbroken sequence into Pusgillian strata and *Diacalymene* Beds of Ashgill age. Here the Ashgill Series is most complete and thickest (600 m) being mainly calcareous mudstones with brachiopods and trilobites (Cautley Mudstones and Ashgill Shales). In the Rawtheyan Stage there occurs the Cautley Volcanic Group, a sequence of contemporaneous rhyolitic tuffs and lavas. The Ashgill sequence in the Lake District is condensed and there are several unconformities. The Applethwaite Beds (Cautleyan), ashy, calcareous mudstones with impure limestone bands, conglomeratic at base, are overlain by 'horny' limestone and a 5-m bed of ash possibly equivalent in part to the Cautley Volcanic Group. The strata above the ash, the *'Phacops' mucronatus* Beds and Ashgill Shales (grey calcareous mudstones) equate with the Hirnantian beds of the Cautley area. In the Cross Fell Inlier, as at Cautley, there was continuous sedimentation from Onnian to Pusgillian and the *Diacalymene* fauna (Cautleyan) is present in the sandy topmost beds of the Dufton Shales. Near Dufton these rocks are unconformably overlain by the Swindale Limestone (Rawtheyan) and Ashgill Shales. Farther south, the 50-m thick, bioclastic Keisley Limestone is also of Ashgill age, but its exact position in the zonal sequence is uncertain. Lithologically similar rocks crop out at intervals along the main Lake District outcrop, and they may represent limestone knolls or reefs, with a fauna distinct from that of the surrounding shales.

3. Silurian System

The Silurian rocks of the district crop out in three distinct areas (Plate XIII). The main area comprises the southern fells of the Lake District, the northern parts of Furness, much of the ground around Kendal, and the Howgill Fells near Sedbergh. The rocks in this outcrop in general form a gently-dipping sequence folded about north-easterly to east–west axes. More acute folding occurs in Furness and major faulting affects the Howgill Fells. The second area is the Cross Fell Inlier, where there are several fault-bounded outcrops, and the third lies along the Scottish border.

The lithostratigraphical classification of the Silurian rocks now current is largely that made familiar by the classic researches of the late Professor Marr and his colleagues, and is set out, together with the biostratigraphical succession, in Fig. 7. The rock-types and facies involved are summarized in Fig. 8. The zonal sequence is based on the recognition of distinct graptolite assemblages and some of the more important zone-fossils are illustrated in Fig. 9. Correlation with Silurian strata elsewhere in Britain, and notably with the type sections of the Llandovery, Wenlock and Ludlow series and stages in South Wales and the Welsh Borderland is established mainly on the basis of the graptolite zones.

Llandovery Series

The Llandovery rocks, collectively known as the Stockdale Shales from their type locality, Stockdale Farm, 4 km east-north-east of Kentmere, Westmorland, are subdivided into the Skelgill Shales below and the Browgill Beds above. They were deposited under conditions of relatively quiet sedimentation, and the Skelgill Shales in particular are generally composed of black, graptolitic mudstone, in which the fossils are commonly preserved in pyrite. The rock is carbonaceous (up to 3·7 per cent carbon) and it contains about 2 per cent sulphur. Interbedded with the graptolitic mudstone are bands of grey unfossiliferous mudstone. One such band—the 'Green Streak' —consists of greyish green mudstone 5 to 10 mm thick, developed within black graptolitic mudstone of the *Monograptus argenteus* Zone, and has been recognized throughout northern England and also widely in central Wales.

At the type locality of the Skelgill Shales—Skelgill Farm, Ambleside, Westmorland—the graptolitic mudstone forms about half of the total thickness of some 18 m. This thickness is almost doubled (33 m) in the Howgill Fells, and although the section at Skelgill may not be complete there is still an appreciable thickening in a north-easterly direction. In the Cross Fell area the Skelgill Shales are found in separated fault blocks, but their thickness is apparently comparable to that in the Howgill Fells.

The Browgill Beds, 90 m thick, have their type locality at Browgill, near Stockdale, and consist mainly of greyish green non-graptolitic mudstone. Some ash bands, graptolitic dark grey mudstones and red mudstones also

SYSTEM	SERIES	STAGE	GRAPTOLITE ZONE	FORMATION (N. ENGLAND)	
DEVONIAN		Downtonian	*No graptolites recorded in Great Britain	Kirkby Moor Flags	
		Whitcliffian			
SILURIAN	Ludlow	Leintwardinian	*Saetograptus leintwardinensis leintwardinensis*	Bannisdale Slates	
		Bringewoodian	*Saetograptus l. incipiens* [or *Pristiograptus tumescens*]		
		Eltonian	*Cucullograptus scanicus* and *Neodiversograptus nilssoni*	Coniston Grits	
				Upper Coldwell Beds	
	Wenlock		*Pristiograptus ludensis* [*Gothograptus nassa* Subzone] *Cyrtograptus lundgreni* *Cyrtograptus ellesae* *Cyrtograptus linnarssoni* *Cyrtograptus rigidus* *Monograptus antennularius* *Monograptus riccartonensis* *Cyrtograptus murchisoni* *Cyrtograptus centrifugus*	Middle Coldwell Beds	
				Lower Coldwell Beds	
				Brathay Flags	
	Llandovery	Telychian	*Monoclimacis crenulata* *Monoclimacis griestoniensis* *Monograptus crispus* *Monograptus turriculatus* [*Rastrites maximus* Subzone] *Monograptus sedgwickii*	Browgill Beds	Stockdale Shales
		Fronian			
		Idwian	*Monograptus convolutus* *Monograptus argenteus* [or *Monograptus leptotheca*] *Diplograptus magnus* *Monograptus triangulatus*	Skelgill Shales	
		Rhuddanian	*Monograptus cyphus* *Monograptus acinaces* *Monograptus atavus* *Akidograptus acuminatus* *Glyptograptus persculptus*		

* Outside Great Britain, graptolites persist into the Lower Devonian. The base of the Downtonian may equate with the base of the *Monograptus ultimus* Zone and the base of the overlying Dittonian with the base of the *Monograptus uniformis* Zone. Recent work suggests that the latter horizon approximates to the base of the Gedinnian, the lowest stage of the European marine Devonian sequence. However, pending ratification of an international agreement that the base of the M. *uniformis* Zone be taken as the base of the Devonian, the boundary is here taken conventionally at the base of the Downtonian.

FIG. 7. *Table summarizing Silurian stratigraphy in northern England*

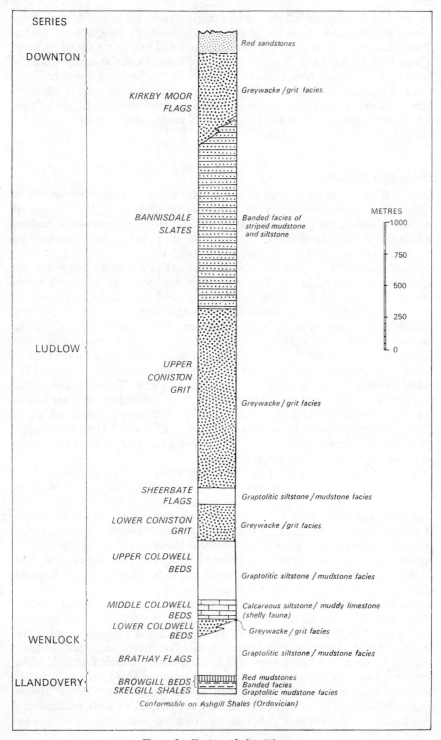

FIG. 8. *Facies of the Silurian*

occur. The last mentioned appear to occupy almost the whole of the *Mono-climacis crenulata* Zone in the Howgill Fells, although the graptolite assemblage characteristic of this zone has been recorded only from Swindale Beck, Cross Fell. In the Lake District the red beds are thinner and impersistently developed, and grey beds immediately underlie the Wenlock Brathay Flags.

In north-east England a small area immediately north-west of Berwick upon Tweed is underlain by greywackes and mudstones of the Upper Llandovery Series. This outcrop is part of a larger one north of the Scottish border.

Wenlock Series

The Wenlock strata of both the main Silurian outcrop and the Cross Fell Inlier consist predominantly of dark bluish grey graptolitic laminated muddy siltstones, with thin siltstone bands, known collectively as the Brathay Flags from their type locality 4 km south-west of Ambleside, Westmorland, together with the Lower and Middle Coldwell Beds.

The Brathay Flags have an average thickness of 300 m and comprise the *Cyrtograptus centrifugus* to *C. lundgreni* zones (see Fig. 7). Small ovoid calcareous concretions are not uncommon, and towards the base of the sequence a 100-mm thick limestone or tabular nodule band occurs within 250 mm of the top of the *C. centrifugus* Zone throughout the Lake District and Howgill Fells, indicating a uniformity of deposition contrasting with that of the Upper Llandovery. The Brathay Flags thicken and coarsen slightly when traced southwards within the Howgill Fells, and westwards from the Howgill Fells to the Lake District. In the *C. lundgreni* Zone this coarsening culminates in the presence of greywackes with thin mudstone partings known as the Lower Coldwell Beds, from their type locality west of Windermere. The greywackes, which here reach a maximum of about 200 m, thin rapidly eastward to be represented by only 12 to 13 m of Brathay Flags with numerous siltstone bands in the Troutbeck area.

The siltstones of the Brathay Flags are believed to be the more distal product of the turbidity currents responsible for the very thick Wenlock greywacke sequence of the south of Scotland and north Northumberland. An inlier extending from Lumsden and Ramshope Burns to Coquet Head, Fulhope, Northumberland, is made up of graded greywackes, siltstones and mudstones with thin graptolitic siltstone bands dipping steeply and in places inverted. The fauna once believed to indicate the presence of both Wenlock and Ludlow rocks is now considered to be solely of Lower Wenlock (*Monograptus riccartonensis* Zone) age.

Detailed petrological studies of the graptolitic mudstones of the Brathay Flags show them to be identical to those in the Wenlock of south Scotland and north Northumberland and in the Wenlock and Ludlow of north and central Wales. They are also essentially similar, except in respect of grain size, to those of the Llandovery and Ludlow of north-west England, which are, however, much finer grained and coarser grained, respectively. In addition there is an upward decrease in carbon content through the Silurian and an increase in evidence of current activity, calcareous material and benthonic fauna. However, despite this detailed knowledge, the question

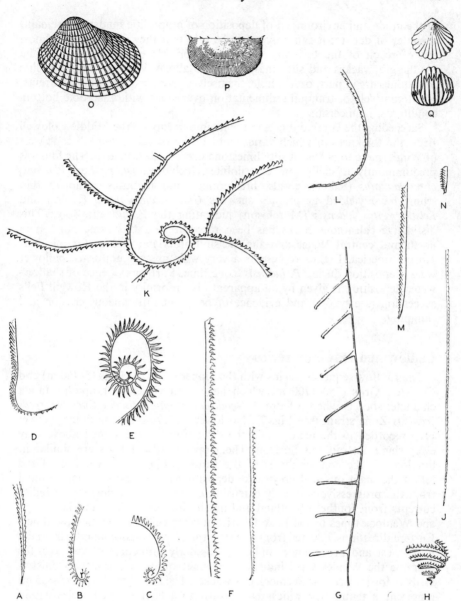

FIG. 9. *Silurian Fossils*
(All natural size.)

Skelgill Beds—**A,** *Akidograptus acuminatus acuminatus* (Nicholson); **B,** *Monograptus triangulatus triangulatus* (Harkness); **C,** *M. triangulatus separatus* Sudbury; **D,** *M. argenteus argenteus* (Nicholson); **E,** *M. convolutus* (Hisinger); **F,** *M. sedgwickii* (Portlock), distal and proximal portions.

Browgill Beds—**G,** *Rastrites maximus* Carruthers; **H,** *M. turriculatus turriculatus* (Barrande); **J,** *Monoclimacis griestoniensis* (Nicol) (s.l.).

Brathay Flags—**K,** *Cyrtograptus murchisoni* Carruthers.

Lower Coniston Grits—**L,** *Neodiversograptus nilssoni* (Barrande) (s.l.).

Bannisdale Slates—**M,** *Pristiograptus tumescens* (Wood); **N,** *Saetograptus leintwardinensis leintwardinensis* (Lapworth).

Coniston Grits—**O,** *Cardiola interrupta* J. de C. Sowerby.

Kirkby Moor Flags—**P,** *Protochonetes ludloviensis* Muir-Wood; **Q,** '*Camarotoechia*' *nucula* (J. de C. Sowerby).

of the mode and environment of deposition of graptolitic mudstones remains a matter of debate; it can only safely be said that they accumulated almost out of reach of the turbidity currents responsible elsewhere at this time for the greywackes and siltstones alluded to above. Whether the graptolitic mudstones are in part, or in whole, themselves products of turbidity currents, or alternatively of tranquil sedimentation over areas with anaerobic bottom conditions, is uncertain.

Succeeding the Brathay Flags in the main outcrop are the Middle Coldwell Beds, the thickness of which varies from 120 m of calcareous siltstone west of Windermere to 5–8 m of silty limestone in the Sedbergh area. The fauna is predominantly a shelly one of trilobites, including *Dalmanites* [*Phacops*] *obtusicaudatus* (Salter), corals, brachiopods and molluscs, although thin seams have yielded graptolites such as *Gothograptus nassa* (Holm) and *Pristiograptus ludensis* (Murchison) indicating the *P. ludensis* Zone. This distinctive calcareous facies has been recognized at this same horizon in north and central Wales and the Welsh Borderland and as far afield as North America. It thus represents a very wide area of uniform, shallower water deposition during *P. ludensis* Zone time. Further evidence of shallow-water deposition is given by an apparent disconformity in the Howgill Fells succession (see below) and evidence of penecontemporaneous erosion and slumping.

Ludlow and Downton series

The Ludlow sequence begins with the Upper Coldwell Beds (5–460 m) and Coniston Grits (750–1700 m) which have both yielded a graptolite fauna characteristic elsewhere of upper *Neodiversograptus nilssoni–Cucullograptus scanicus* Zone strata (see Fig. 7). No lower *N. nilssoni* Zone fauna has yet been recorded in the region; and the disconformity mentioned above may everywhere cut out this horizon. The Upper Coldwell Beds are similar in lithology to the coarser facies of the Brathay Flags. The Coniston Grits reflect the main onset of greywacke deposition in the region which resulted from the progressive southerly extension of greywacke-depositing turbidity currents from southern Scotland and north Northumberland in Llandovery and Wenlock times to the Lake District and Howgill Fells in Ludlow times. Current directions, inferred from sole structures, were predominantly from the north-west and provenance studies, particularly on coarse greywackes, for example the Winder Grit, indicate the same source area for the Coniston Grits as for the Scottish Wenlock greywackes. The Sheerbate Flags (75–355 m) represent a temporary widespread return to a finer grained facies within the Coniston Grits sequence.

The succeeding Bannisdale Slates, which average 1540 m in thickness, comprise a banded facies consisting of dark grey mudstone with stripes of pale siltstone believed to be, like the greywackes, the products of turbidity currents. Sparsely graptolitic mudstone bands contain *Saetograptus leintwardinensis incipiens* Zone and *S. leintwardinensis leintwardinensis* Zone forms. The overlying Kirkby Moor Flags (450–900 m) consist of well-sorted sandstones and siltstones, often slumped, with lenses of fossils. They have an entirely shelly fauna (Fig. 9) and, mainly on the basis of ostracods, are believed to be of Upper Ludlow and basal Downtonian age. The highest member of the formation, the Scout Hill Flags (type locality 11 km south-south-east of Kendal) consists of grey and red current-bedded siltstones and sandstones. It represents the final stage of infilling of a basin under the increasing influence of south-easterly flowing currents. The basin probably subsided faster than the rate of sedimentation during Lower Llandovery time and at a progressively slower rate during the Wenlock and Ludlow, until the basin was ultimately filled up during Downtonian time.

Local facies variation throughout the post-Wenlock succession is marked and has led to the recognition locally of such formations as the Longsleddale Siltstone (0–120 m) in the Coniston Grits and Lower and Upper Underbarrow Flags (0–240 m) at the base of the Kirkby Moor Flags. There is evidence of axially directed currents in a north-east to east–west trending basin and these probably account for much of this variation and that, in thickness and grain size, noted in the Upper Llandovery and Wenlock (Brathay Flags and Lower Coldwell Beds).

4. Devonian System: Caledonian tectonics and igneous activity

Earth movements

Throughout the Lower Palaeozoic epoch the north of England was an area of considerable tectonic activity. Periods of relative stability and quiet deposition alternated with periods of rapid subsidence and sedimentation under geosynclinal conditions, during which thousands of metres of rock accumulated. Minor breaks occur throughout the sequence, and during three of these intervals folding and uplift sufficient to produce marked unconformities took place. Of two of these episodes, both of Ordovician age, one preceded and one succeeded the outpouring of the tuffs and lavas of the Borrowdale Volcanic Group. The third, the main Caledonian orogeny, postdates the Kirkby Moor Flags (Downtonian) in the Lake District, and antedates the andesitic lavas of the Cheviots, of presumed Lower Devonian age, which unconformably overlie folded Silurian strata. During these episodes, the rocks involved were in places highly cleaved and contorted. Whilst the predominant trend of the structures produced is Caledonoid (north-east to south-west), cross-folding also occurred, and at some localities three or more cleavages can be seen in one hand specimen.

Igneous rocks of the Cheviots

The igneous complex of the Cheviot Hills consists of the deeply dissected remains of a volcano intruded by granite and traversed by igneous dykes. Since it rests on upturned Silurian strata and, on the Scottish side of the Border, is transgressed by Upper Old Red Sandstone conglomerates wherein fragments of the lavas and granite abound, the age of the complex is regarded as Lower Devonian.

Amongst the lavas there are rare deposits of red sandstone or marl. These, the only sedimentary rocks, are unfossiliferous, and so throw no direct light on the age of the complex.

Apart from certain small patches to be noted later (Chapter 5) this volcanic pile is the only representative in the region of beds of possible Lower Devonian age. Their absence from the Lake District, the Alston Block and the southern margin of the Southern Uplands may be due to non-deposition, but could equally well be the result of erosion following mid-Devonian folding and uplift; and thick deposits of this age may well underlie the Northumbrian Trough.

In the Cheviots, the first stages of vulcanicity were explosive, and produced coarse ashes and agglomerates which rest upon folded and eroded Silurian rocks. This episode was quickly followed by an immense out-pouring of lavas which even today after extensive denudation cover an area of 600 sq km. Into this volcanic pile a large mass of granite was intruded, which now appears as a core of plutonic rock surrounded by the lavas and tuffs which dip gently away from it. In the concluding stages of this cycle of igneous activity the intrusion of a series of dykes cutting both lavas and granite was followed by extensive pneumatolytic alteration, including tourmalinization.

There are surprisingly few exposures on the steep, rounded hills of the Cheviots, but conditions are better in the valleys, save where—as is often the case—the stream courses have been determined by crush lines. It is not easy, therefore, to make out a detailed sequence. The Basal Agglomerate, 60 m thick at most, is known only on the south side of the Cheviots, close to the Scottish border. The included fragments, often large and even up to 1.5 m across, are usually fine-grained mica-felsites or rhyolites: they resemble the rocks of certain flows in the succeeding group, though some are of a type not yet known *in situ*; others, smaller and near the base, are of Silurian shale.

Succeeding the Basal Agglomerate, and with a similarly restricted distribution, is a suite of porphyritic rhyolites sometimes known as the Biotite Lavas or 'Mica-Felsites'. These are purple, brick-red or chocolate-coloured, and both feldspar and mica phenocrysts are prominent. The remaining lavas —the Pyroxene Andesites—make up the greater part of the Cheviot Massif. They are mainly purple or dark brown in colour except within a kilometre or so of the granite, where they are dark grey, a change possibly due to contact alteration by the granite. Although the vast majority are augite-hypersthene-andesites, other types are known: these include glassy or pitchstone varieties near the Border to the west, and flows of trachyandesite to the south-east; in the latter quarter, beds of ash associated with the lavas attain their greatest development; elsewhere they are of less account but are not uncommon. The lavas are frequently scoriaceous and amygdaloidal, and the cavities are commonly filled with chalcedony.

The granite outcrop forms a roughly circular area some 52 sq km in extent in the centre of the lavas. The upper surface of the granite appears to be dome-shaped and, in places, remnants of the hornfelsed lavas forming part of the original roof are to be seen. The intrusion is variable in composition, the marginal and central parts being of diopside-granodiorite, with augite xenocrysts, separated by an intervening zone of pink pyroxene-free granophyre. There is a well-defined junction with the lavas on the south, but on the north granite and lavas are intricately inter-mixed, crushed and brecciated.

Aplitic veins traverse both the granite and the surrounding lavas, and in their turn are cut by dykes. The latter usually trend either north-north-west or north-north-east and are of four main types, of which mica-porphyrite is the commonest, while the rest comprise augite-hypersthene-porphyrites, quartz-porphyrites and felsites.

Intrusions in the Alston Block

On the Alston Block, Lower Palaeozoic rocks are exposed in the Teesdale and Cross Fell inliers. They are intersected by numerous intrusions of dolerite, lamprophyre and acid porphyry. There is insufficient space to detail them here, but special mention may be made of the Cuns Fell Dolerite, cutting rocks of the Skiddaw Group west of Cross Fell, and of the Dufton Microgranite, a quartz-feldspar-mica-porphyry intruded in Dufton Shales, near the village of Dufton.

Weardale Granite

Gravity-field studies on the Alston Block have suggested that it is underlain by a large granitic intrusion, shown schematically in Fig. 2. This rock is nowhere exposed at the surface, but its existence has been proved by the Rookhope Borehole (Dunham and others 1965). The Carboniferous rocks in the bore rest unconformably on an eroded surface of the granite, showing it to be pre-Carboniferous in age. Radiometric dating has since established a mean age of about 364 m.y. for the intrusion, which is a biotite-muscovite-granite, cut by numerous small veins of aplite and pegmatite.

Intrusive rocks of the Lake District

Probably no district in England of equal size can furnish such a variety of intrusive rocks as the Lake District. Many of them are acid in character, i.e. light-coloured igneous rocks, such as granites, containing much free quartz, but there are also examples of intermediate, basic and ultra-basic rocks; these differ from acid rocks in being, as a rule, darker in colour, and in possessing little or no free quartz but more of the ferromagnesian minerals —biotite, hornblende, augite or olivine. The intrusions also show consider-able variations in form—stocks (large, steep-sided masses), bosses (irregular), laccoliths (lenticular), dykes (wall-like) and sills (sheet-like) probably all being represented.

Although confined to Ordovician rocks, most of these intrusions have generally been assumed to be of Devonian age; and this has been confirmed in several cases by radiometric age determinations. The more important of these intrusions will be dealt with in anti-clockwise sequence, according to their positions on Fig. 4.

Shap Granite

The outcrop of this granite is about 8 sq km in extent and forms a rough oval extending east and west. According to Marr the intrusion forms a 'cedar-tree laccolith' in the Borrowdale Volcanic Group near the junction with the Coniston Limestone Group, and consists of a suite of porphyritic biotite-granites with dioritic affinities. The content of accessory minerals is high, but special characteristics of this much-used ornamental stone are the large pink crystals of feldspar and the number of inclusions which usually show up as dark patches. These clots, known as 'heathen' are commonly of basic character; many probably represent caught-up relics of an earlier intrusion, but some are partly assimilated fragments of Borrowdale or Coniston Limestone rocks. The 'Light' and 'Dark' Shap Granites of the building trade really grade into each other; they are believed to be due to the degree of pneumatolysis (the effects of heated vapours) on the plagioclase feldspars of the groundmass after consolidation, for the 'dark' variety occurs as ribs 0·5 to 5 m wide on each side of the master joints, and the width of dark rock appears to depend on the size and persistence of a partic-ular joint. The large pink crystals of orthoclase are apparently unaltered. This pneumatolysis is also responsible for the introduction of new minerals, especially molybdenite, into the country rocks.

The last stage of granite intrusion was followed by the injection of a series of minor intrusions including lamprophyres and quartz-felsites, some

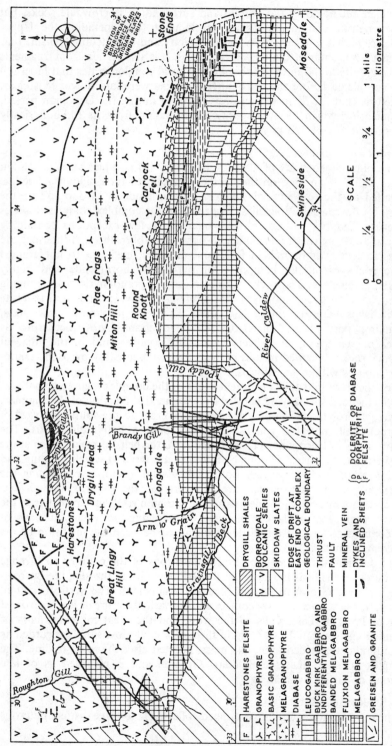

FIG. 10. *Sketch-map of the Carrock Fell Complex*

of which cut not only the granite and surrounding Borrowdale rocks, but also penetrate and metamorphose strata up to and including the Upper Ludlow, and it is clear, from the inclusions and the metamorphism, that intrusion was completed after Silurian times. It is also certain from fragments of the granite found in the Carboniferous basal conglomerates that intrusion was completed before the Carboniferous rocks were laid down. It is clear, therefore, that the Shap Granite was intruded during the Devonian period, a conclusion which receives specific confirmation from its radiometrically determined age of 393 m.y.

Carrock Fell Complex

The plutonic complex of Carrock Fell (Fig. 10) occupies an area 1·6 km broad and 6·4 km long on the northern margin of the Skiddaw Anticline, bounded on the south mainly by rocks of the Skiddaw Group and on the north by Borrowdale Volcanics. The rocks of the complex comprise a series of gabbroic types without olivine, but with a few exceptions containing quartz, and they range from ilmenite-rich melagabbro through intermediate types to leucogabbro rich in feldspars; diabases, medium- to fine-grained basic rocks essentially composed of pyroxenes, hornblende and highly altered plagioclase; acid rocks, which are largely granophyric types, ranging from hedenbergite-granophyre to albite-granophyre; and quartz-felsite which is probably of rather later date. There are also minor intrusions, including porphyrites, quartz-dolerites and granophyres, occurring as dykes or steeply dipping sheets within the complex and in the adjacent country rocks. Included masses of lavas are represented by pyroxene granulite xenoliths within the gabbros.

For the most part the constituent rocks of the complex seem to be narrow, steeply inclined sheets lying parallel to the general elongation of the complex. In the gabbros the variations, as seen in the crags to the south and west of Carrock Fell, reveal a rough symmetry about a longitudinal axis. The central mass of leucogabbro is separated from the belts of melagabbro at the southern and northern margins by intermediate types in which banding and fluxion texture are displayed. Both diabase and granophyre, forming the northern part of the complex, are younger than the gabbros, though it is not certain which was intruded first. The intimate relationships between the various rock types of the complex and the mingling that occurs at many of the contacts all suggest that little time elapsed between the intrusion of the different elements.

On the margins of the complex, the horizons of the country rock range from arenaceous beds low down in the Skiddaw Group through the Borrowdale Volcanic Group to Drygill Shales (Caradocian). The vertical, and in places, transgressive intrusive contact on the south, the more or less horizontal roof of lavas in Thief Gills to the west, and the great marginal mass of included lavas at Snailshell Crag in the east are all most simply regarded as bounding a steep-sided, plug-like mass, injected after the main folding. The complex is older than the Skiddaw Granite, and is regarded as of late Silurian or early Devonian age.

The Skiddaw Granite

The term Skiddaw Granite is applied collectively to three distinct outcrops of white or grey biotite-granite that lie within a region of thermally altered slates of the Skiddaw Group (Fig. 11). Their petrographic similarity and the form of the thermal aureole suggest that they belong to one granite mass and are connected at no great depth below the surface. The largest area is referred to as the central, or Caldew outcrop; the northern one as the Grainsgill outcrop; and the third and smallest as the Sinen Gill outcrop.

The thermal aureole of the granite has the form of a broad flat-topped dome. The early stages of metamorphism of the Skiddaw Slates are indicated by the development of spotting and of porphyroblasts of chiastolite. With increasing metamorphism there is a progressive hardening of the rock to a massive hornfels, the spots become clearly defined crystals of cordierite, the

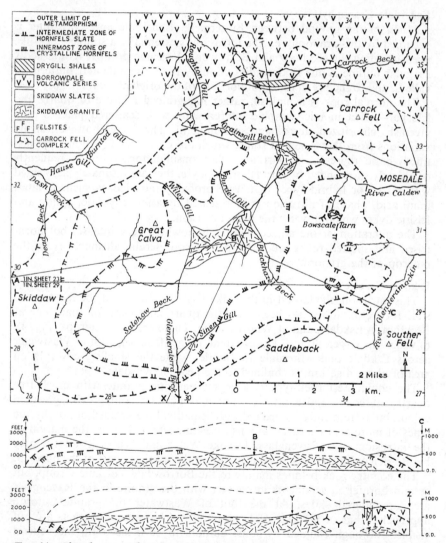

FIG. 11. *Sketch-map and sections showing the metamorphic aureole of Skiddaw Granite*

chiastolite changes to clear translucent andalusite and biotite becomes a visible constituent of the base. The alteration also affects adjacent parts of the Carrock Fell Complex.

In the Grainsgill area, late-stage metasomatism has altered much of the normal biotite-granite to greisen, mainly a quartz-muscovite rock with small amounts of accessory minerals. Close to the greisen, the normally black, hornfelsed slates are extensively altered to pale grey rocks in which muscovite is conspicuous.

On structural grounds, a late Silurian or early Devonian age is indicated for the Skiddaw Granite, and this is confirmed by radiometric age determinations which give a mean age of 399 m.y.

Ennerdale Granophyre

Much of the ground between Buttermere, Ennerdale and Wastwater is occupied by this granophyre. The intrusion is essentially a simple stock-like mass invading the Skiddaw and Borrowdale Volcanic groups; its markedly irregular outline is due to deep dissection by the Ennerdale and Buttermere valleys, and to the preservation of portions of the original roof.

The normal rock is a pinkish, rather fine-grained mixture of quartz and feldspar—often in micrographic intergrowth—with scattered patches of a greenish, chloritized, ferromagnesian mineral. There are, however, many variations from this type and some, which are dark grey in colour and distinctly basic, represent, in part, an earlier consolidation; although frequently completely surrounded by normal granophyre, they are intimately associated with mixed or hybrid types that were probably intruded as such. A later, more acid, phase of intrusion is represented by a few rhyolitic, aplitic and felsitic dykes which traverse the main intrusion.

The alteration of the adjacent rocks is nowhere very intense, but hornfelsed and spotted rocks are developed at considerable distances from the outcrop of the granophyre.

Eskdale Granite

This intrusion, the largest in the Lake District and one of the most accessible for exploitation, has been but little quarried, except for roadstone at Beckfoot in Eskdale, and at Waberthwaite. There are two outcrops (Fig. 4), of which the larger, 20 by 7 km in extent, ranges from the foot of Wastwater across Eskdale and Miterdale to Bootle, while the smaller, covering an area of about 4 sq km, is confined to the head of Wasdale.

The commonest rock-type is a coarse perthitic granite with much free quartz, muscovite and a little biotite; it is pink or light greyish green in colour, but is frequently stained red with hematite. Fine-grained or porphyritic types of more acid character are also present, especially near the margin of the intrusion near Ravenglass, Devoke Water and Waberthwaite; a grey biotite-rich granite is developed throughout the southern part of the outcrop.

The country rock belongs mainly to the Borrowdale Volcanic Group, but certain hornfelsed sediments in contact with the granite in the Ravenglass area are altered slates. At the foot of Wastwater the granite outcrop approaches that of the Ennerdale Granophyre, but actual contacts are not visible. Although claimed by some authors to be a laccolith, the general shape of this intrusion conforms more to that of a stock.

The mean radiometric age of the Eskdale granite is 383 m.y., suggesting that it also is of Devonian age.

Other Intrusions

Space does not permit of the description of the many other intrusions in the Lake District (Fig. 4), but attention may be drawn to the Threlkeld Microgranite, which has been intruded between the Skiddaw and Borrowdale Volcanic groups. Although the mass is of considerable extent the amount of metamorphism is negligible, and graptolites have been found within 0·5 m of the contact. There is also the so-called 'Embleton Granite' in the Skiddaw Slates between Bassenthwaite and Cockermouth; this is a quartz-mica-diorite with strong affinities to the granophyres. Dolerite occurs at Castle Head—which has been claimed to be a volcanic neck—and at Friars Crag, Keswick; while the minette of Sale Fell, Wythop, the picrite at Dash near Skiddaw, and the granophyric Armboth Dyke in the Thirlmere area are worthy of mention. Farther east there is the Haweswater Dolerite Complex which, as noted previously (p. 16), may be of Ordovician age.

Intrusive rocks in the Manx Slates

The rocks included under this heading (Fig. 12) are confined to the Manx Slates into which they were intruded before Carboniferous times. Dykes of Carboniferous and Tertiary age are dealt with elsewhere.

There are two major intrusions of granite in the island, at Dhoon and at Foxdale; they differ considerably in mineralogical composition and in mode of occurrence. That of Dhoon is a porphyritic microgranite and appears to form a rude stock with steep walls, whereas the Foxdale mass, which is a grey coarse muscovite-granite, shelves gradually beneath the slates and is possibly a laccolith. Two smaller plutonic intrusions may be mentioned. That near Oatland consists of gabbroic rocks pierced by a granite of Dhoon type which has assimilated some of the basic rocks invaded; that of Ballabunt is a quartz-diorite.

Minor intrusions of pre-Carboniferous age are represented by two series of dykes differing in composition but possessing a uniform north-east to south-west trend along the strike of the Manx Slates. One series, the earlier, consists of microgranites and felsites (elvans) and is definitely related to the granites, being restricted to narrow belts between and on their flanks. The basic 'greenstone' dykes, chiefly dolerites and the lamprophyres of minette type, that characterize the other series, are more numerous and widespread; they have no obvious connection with the granites.

The Manx Slates were folded and crushed previous to the intrusion of granite and of both series of dykes. It is equally clear, however, that they subsequently suffered further lateral compression, for the Dhoon Granite in particular has undergone considerable shearing movement along its margin, and in most cases the dykes have been crushed and dragged out into disconnected lenticular masses.

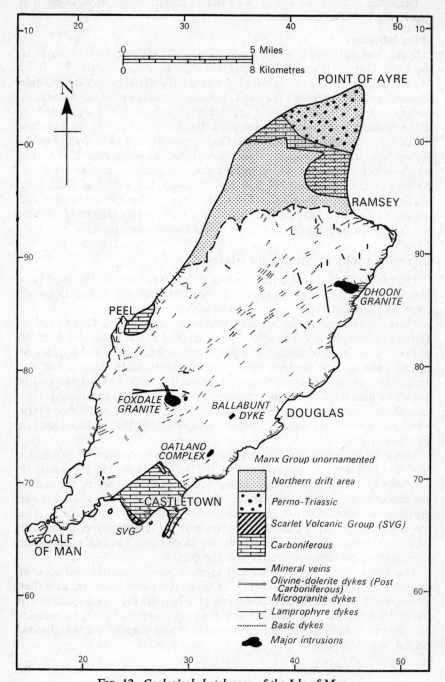

FIG. 12. *Geological sketch-map of the Isle of Man*

5. Carboniferous System

Carboniferous rocks underlie about three-quarters of the region, either at outcrop or concealed by later beds, and as a result of centuries of exploration for minerals, particularly coal and vein ores, we have a far greater understanding of this system than of any other. The main divisions are as follows:

Upper Carboniferous $\left\{\begin{array}{l}\text{Coal Measures (Westphalian)}\\ \text{Millstone Grit Series (Namurian)}\end{array}\right.$

Lower Carboniferous   Carboniferous Limestone Series (Dinantian)

The three divisions on the right of the table, which equate with the main subdivisions in this chapter, are based on lithological characteristics that are particularly well marked in Lancashire and the southern Pennines, where massive marine limestones are succeeded by thick grits that are followed by coal-bearing rocks. In northern England the differences between the two lower divisions are less pronounced; there is much sandstone in parts of the Carboniferous Limestone sequence and limestones are abundant in the Millstone Grit, in which thick gritstones are much less a feature than they are farther south. Nevertheless, it has been found possible with the aid of fossils to correlate the rocks in the two areas, and the same names are applied to the divisions in the interests of nation-wide uniformity. For wider correlations with other parts of the world the names in brackets alongside them are more appropriate, as these are precisely defined in palaeontological terms.

Cyclic Sedimentation

The main Carboniferous rock types are limestone, mudstone, sandstone, seatearth and coal, which are arranged in various cyclic or rhythmic sequences. Each cycle of sediment, or cyclothem, began with an abrupt change in relative sea level giving marine or near-marine conditions. The water shallowed as sediments built up to water-level and ended with the establishment of coal-forests on the newly formed land. Cyclothems are rarely complete, and traced laterally die out or split into two or more. Tournaisian cycles (see p. 39) are wholly marine and without coals.

Probably the best illustration of a complete cyclic sequence is the 'Yoredale' type as shown by Upper *Dibunophyllum* age strata on the Alston Block. The lowest member of each cycle is a marine limestone between 2 and 20 m thick, which is followed upwards by calcareous fossiliferous shale, ferruginous shale with a few marine fossils, silty shales and shaly sandstones, sandstone, seatearth or ganister and finally coal. There are sometimes several minor cycles towards the top with as many seatearths. Coals are rarely more than 150 mm thick. The sandstones are generally cross bedded and may have an erosive base. Cyclothems are generally between 15 and 30 m thick. It is a characteristic feature that the upper terrigenous part, deposited in deltaic conditions, is succeeded abruptly by the limestone at the base of the next cycle.

The cyclicity persisted throughout the Carboniferous and there was a progressive tendency in Millstone Grit times, probably because of the changing balance between the amount of subsidence and the provenance of sediments, for the duration of the marine periods to become shorter, and the resultant limestones thinner. In Coal Measures times truly marine invasions occurred in only small numbers of the many cyclothems, and the periods of coal-formation were longer. The result was a thick series of deltaic strata containing many coals of workable thickness, with a few thin widespread layers of fossiliferous marine mud which are known as marine bands, and which have proved to be of great value in Coal Measures correlation.

Carboniferous Limestone Series (Dinantian)

The Caledonian orogeny and its accompanying erosion created throughout the Northern England region a new basement surface on which sedimentation recommenced in late Devonian or early Carboniferous times. This basement was made up of structurally contrasting areas of 'blocks' or 'massifs' and 'troughs' or 'basins'. The 'blocks' were relatively stable areas, which underwent only very gentle folding in post-Carboniferous times; they were not

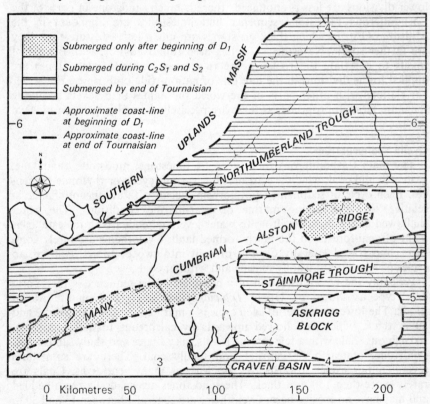

FIG. 13. *Lower Carboniferous geography*

submerged by the sea until after the start of the Viséan (see Fig. 13), and subsided more slowly than the 'troughs', thereafter resulting in relatively thin Carboniferous sequences. By contrast, sedimentation started earlier in the 'troughs', certainly in the Tournaisian and possibly in places earlier; here subsidence and accumulation were rapid and gave vast thicknesses of strata. However, despite the variations in subsidence, sedimentation kept pace with it virtually everywhere, for it is clear that all the Carboniferous sediments were formed near sea level. The depositional surface at any given time across both 'blocks' and 'troughs' alike was almost horizontal, and the sea in which the shelly limestones and shales accumulated was probably a few tens of metres deep at most, while the coals and associated seatearths, together with some of the deltaic sandstones, represent emergences of a few metres.

We have thus the picture of a region which at the beginning of upper Devonian time had a moderate relief with a broad pattern of high and low ground reflecting the 'block' and 'trough' structure. This relief was greatly reduced and perhaps partly buried by the end of the Devonian, and the region then subsided to be progressively drowned by the sea during Lower Carboniferous time.

'Block' margins varied in the sharpness of their definition, both in place and time. They may be marked by extensive faults, some of which appear to have been formed contemporaneously with the sedimentation, and some of which formed much later. The rocks overlying the margins in places change sharply in facies and tectonic style; while elsewhere transitions are gentle. Limestone reefs in the Lower Carboniferous seas occurred only along the margins of the 'blocks'.

The varied tectonic setting and palaeogeography of the region, outlined above, led to variations in the character or facies of the Lower Carboniferous sediments. Most of the terrigenous material came from the north or north-east, so that there was a general tendency for the proportion of limestone to diminish in that direction, though actual thickness variations are more closely related to the distribution of 'troughs' and 'blocks'.

Variations in thickness and facies are particularly marked in Viséan strata, and are shown in Figs. 17 and 18. The division into various facies is arbitrary, for they grade laterally into each other. Most of the names are self-explanatory. The Yoredale facies is described on p. 37. The Northumbrian facies resembles the Yoredale but is less calcareous and more terrigenous, with the cycles thinner, more numerous and not so well defined. Limestones in the Northumbrian facies are thin and tend to be argillaceous; coals are generally thin and seatearths numerous; and the same bed of shale or sandstone may contain both marine fossils and rootlets.

Classification

The Carboniferous Limestone Series in northern England is divided into the following stages and zones:

Viséan Stage:	Upper *Dibunophyllum* Zone (D_2)
	Lower *Dibunophyllum* Zone (D_1)
	Seminula Zone (S_2)
	Upper *Caninia* Zone (C_2S_1)
Tournaisian Stage:	Not divided in northern England but including Lower *Caninia* (C_1) and lower zones.

Zonal divisions in this region were first established by Garwood (1913) in Ravenstonedale and have since been extended, with modifications, over the whole region as follows:

Garwood's zones	Garwood's subzones	Current equivalents
Dibunophyllum	{ *Dibunophyllum muirheadi*	D_2 and lowest Namurian
	Lonsdaleia floriformis	D_2
	Cyathophyllum murchisoni	D_1
Productus corrugato-hemisphericus	{ *Nematophyllum minus*	S_2
	Cyrtina carbonaria	S_2
	Gastropod Beds	C_2S_1
Michelinia grandis	{ *Chonetes carinata*	C_2S_1
	Camarophoria isorhyncha	C_2S_1
Athyris glabristria	{ *Seminula gregaria*	C_2S_1
	Solenopora	} C_1 or lower
	Pinskey Gill Beds	

Facies changes, and in Northumberland lack of diagnostic fossils, have created difficulties in correlation, which are not yet wholly resolved, and different sequences of groups and formations have been set up for different areas (Fig. 14). It is not surprising, therefore, that the extension of some group and formation names, with miscorrelation and imprecision of definition, has led to a certain amount of confusion in the literature. The classifications currently in use are shown in Figs. 14, 16 and Plate V, and will be used in the rest of this chapter without further definition.

Basal Conglomerates

At various places (see Fig. 14) there are outcrops of conglomerates and sandstones resting with strong unconformity on Silurian or older rocks, and overlain by marine Tournaisian or Viséan sediments. At present, these conglomerates can only be dated as post-Silurian and older than the overlying marine sediments, but some, at least, may span the Devonian–Carboniferous time boundary.

Perhaps the oldest is in the Mell Fell area north of Ullswater, where red conglomerates and coarse-grained sandstones, about 1500 m in total thickness, form a fan-like deposit of locally derived materials. Quartz-conglomerates and sandstones beneath Viséan sediments crop out in places along the Pennine escarpment north of Roman Fell, where they attain 200 to 250 m. Farther south there are outliers of red and brown conglomerates and sandstones near Sedbergh, north of Kirkby Lonsdale and at Barbon, resting unconformably on an uneven surface of Silurian rocks, and considered to be piedmont fans formed in a rather arid climate.

Near Peel, on the west coast of the Isle of Man, the Peel Sandstones form an outlier with an estimated thickness of some 400 m on the Manx Slates. They consist of red sandstones with partings of red shale, beds of conglomerate with pebbles of quartz, sandstone and Silurian limestone, and lenticular impure concretionary limestones.

Area:	SOUTH ISLE OF MAN	FURNESS	WEST CUMBERLAND	ALSTON BLOCK	BROUGH-RAVENSTONEDALE	NORTH CUMBERLAND	W & NW NORTH-UMBERLAND	NORTH NORTH-UMBERLAND
Overlying beds								
P2 / D2	SCARLET VOLCANIC GROUP *; BLACK LIMESTONES	GLEASTON GROUP (*Girvanella Band*)	HENSINGHAM GROUP (*First Lst*)	UPPER LIMESTONE GROUP (*Great Lst*)	MILLSTONE GRIT SERIES (*Great Lst*)	MILLSTONE GRIT SERIES (*Great Lst*)	UPPER LIMESTONE GROUP (*Great Lst*)	UPPER LIMESTONE GROUP (*Great-Dryburn Lst*)
P1 / D2	POYLLVAAISH LIMESTONES			MIDDLE LIMESTONE GROUP (*Smiddy Lst*)		UPPER LIDDESDALE GROUP (*Low Tipalt Lst*)	MIDDLE LIMESTONE GROUP (*Oxford Lst*)	MIDDLE LIMESTONE GROUP (*Oxford Lst*)
B2 / D1		URSWICK LIMESTONE	CHIEF LIMESTONE GROUP		ALSTON GROUP	*Naworth Bryozoa Band*	*Lower Bankhouses Lst*	LOWER LIMESTONE GROUP
D1	CASTLETOWN LIMESTONES			LOWER LIMESTONE GROUP (*Melmerby Scar Lst*)	*Great Scar Lst*	LOWER LIDDESDALE GROUP	LOWER LIMESTONE GROUP (*Redesdale Lst*)	(*Dun Lst*)
S2		PARK LIMESTONE	*Seventh Lst*					
S2			COCKERMOUTH LAVAS; BASAL CONGLOMERATE *	SHALES WITH LIMESTONE	ORTON GROUP	UPPER BORDER GROUP (*Clattering Band*)	SCREMERSTON COAL GROUP	SCREMERSTON COAL GROUP
C2S1	BASAL CONGLOMERATE	DALTON BEDS; RED HILL OOLITE; MARTIN LIMESTONE; BASEMENT BEDS *				MIDDLE BORDER GROUP (*Whitberry Band*)	FELL SANDSTONE GROUP †	FELL SANDSTONE GROUP †
Strata not divided into zones				BASEMENT BEDS *	*Algal Layer*; RAVENSTONEDALE GROUP; *Pinskey Gill Beds*	CAMBECK BEDS; MAIN ALGAL BEDS; BEWCASTLE BEDS; LYNEBANK BEDS; LOWER BORDER GROUP; *base not seen*	CEMENTSTONE GROUP; COTTONSHOPE LAVAS *; LOWER FREESTONE BEDS *	CEMENTSTONE GROUP; KELSO LAVAS *

LOWER CARBONIFEROUS or DINANTIAN — VISEAN / TOURNAISIAN

Fig. 14. *Table summarizing classifications of Lower Carboniferous stratigraphy in northern England.* Oblique ruling is used where beds of the indicated age are not represented.

* age uncertain
† both top and bottom of the Fell Sandstone Group are diachronous

Other conglomerates at the base of Lower Carboniferous sequences are known in the north and south of the Isle of Man, west Cumberland, Furness, on the Alston Block and near the Scottish border at Ramshope, Windy Gyle and Roddam. These are all referred to below.

Tournaisian Rocks

The distribution of the marine Tournaisian strata in northern England is shown on Fig. 15 which also shows their relationship to the corresponding rocks on the Scottish side of the Border. In the following account details of these occurrences are set out geographically from south to north.

Ravenstonedale. The Ravenstonedale Group in its type area begins with the Pinskey Gill Beds (Fig. 16). These rest unconformably on Silurian strata, and comprise shales, sandstones and dolomitic limestones, with a very restricted fauna including the brachiopod *Spirifer pinskeyensis* Garwood. Their relationships to the red sandstones and conglomerates of the Birk Beck Valley, south of Shap, are uncertain, but they both appear to be overlain by the Feldspathic Conglomerate, a sequence of sandstones and shales with much granitic detritus. The limestones which follow, up to the Algal Layer, are presumed to be of Tournaisian age. The fauna, characterized by '*Camarotoechia*' *proava* (Phillips) and nodules of the alga *Solenopora*, does not closely resemble that of corresponding strata in either the Northumberland Trough to the north or the Craven basin to the south; and the lithology of the beds—thinly bedded dolomitic limestone with calcite-filled vugs—lends support to the suggestion that these rocks may have been deposited in a restricted embayment, possibly with marine access only eastwards into the Stainmore Trough. The *Spongiostroma* and algal beds at the top of the sequence are the result of very shallow-water deposition, and may have been in part interor even supra-tidal in origin.

Alston Block. Tournaisian strata crop out near the south-west corner of the Alston Block on and near Roman Fell (Fig. 15). Here the Basal Conglomerate is overlain by the Roman Fell Shales and the Roman Fell Sandstones. Thicknesses either side of the west-north-westerly trending Swindale Beck Fault differ and indicate contemporaneous movement along the fault, which persisted into Viséan times. The Basal Conglomerate contains locally derived pebbles from the Lower Palaeozoic rocks on which it lies with an uneven base. The Roman Fell Shales are mainly purple and green silty mudstones, some 45 m thick north of the Swindale Beck Fault and 60 m thick south of the fault. The overlying Roman Fell Sandstones, which are sandstones and conglomerates with many quartz and mudstone pebbles, increase in thickness across the fault from 60 to 180 m on the south side. The Roman Fell Beds on the west face of the fell are stained purplish red in common with the underlying Lower Palaeozoic strata, but eastwards they become yellowish.

Bewcastle. About 700 m of Tournaisian sediments are exposed in the Bewcastle Anticline of north Cumberland, where they comprise the three lower divisions of the Lower Border Group (Figs. 14 and 16). The base is not seen, but comparison with the sequence in Liddesdale suggests that it lies not far below surface. All the strata show rhythmic sequences of mudstones and limestones with subordinate sandstones, deposited in a shallow

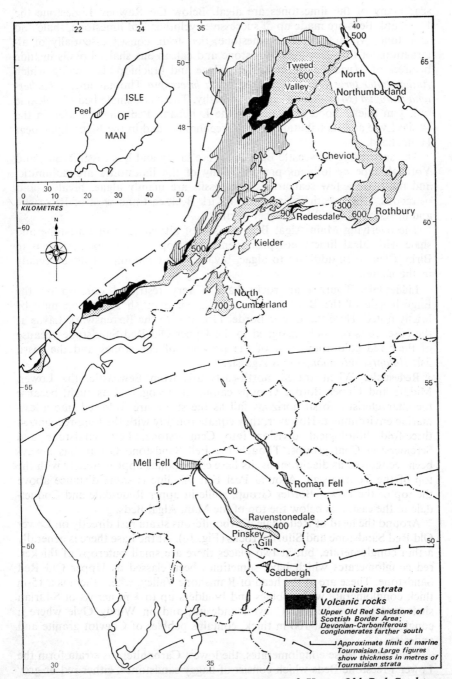

FIG. 15. *Distribution of outcrops of Tournaisian and Upper Old Red Sandstone rocks in northern England and the Scottish border*

sea; many of the limestones are algal. Below the Rawney Limestone the Lynebank Beds are made up of successive cyclothems of limestone, shale and sandstone, while above this limestone, the strata consist essentially of alternations of thin dolomitic limestone and calcareous shale. Fossils include bivalves of *Myalina*-type, ostracods, algae and brachiopods, among which *Antiquatonia teres* (Muir-Wood), *Pustula interrupta* Thomas and *Schuchertella ambigua* (Muir-Wood) are noteworthy. There is a single bed of volcanic tuff, $\frac{1}{2}$ m thick. Good exposures of the Lynebank Beds may be seen in the White Lyne and its tributaries near Greenholme, and in the Black Lyne near Holmehead.

The succeeding Bewcastle Beds are more sandy and less fossiliferous, and Yoredale-type cyclothems prevail. Many of the limestones are dolomitic, and there are a few seatearths. The fossils are mainly algae, bivalves and ostracods. Good sections are seen in Kirk Beck east and west of Bewcastle and in Ashy Cleugh 2 km farther north.

The overlying Main Algal Beds consist of alternations of limestone and shale with algal limestones prominently developed. The type section is in Birky Cleugh. In addition to algae, bivalves are common fossils, especially in the shales.

Liddesdale. Tournaisian rocks have a very restricted outcrop on the English side of the Border, but the adjacent Scottish succession may be briefly noted. Here the Lower Border Group with the Birrenswark Lavas at the base, rests on rocks assigned to the Upper Old Red Sandstone because of their red coloration, and of the presence of cornstones and the fossil fish *Holoptychius nobilissimus* Agassiz.

Redesdale. When traced north-eastwards from Bewcastle, the Lower, Middle and Upper Border Groups cannot be recognized as such, because the characteristic fossil horizons fail as the strata are followed into a less-marine environment. However, they equate roughly with the long-recognized three-fold lithological division into Cementstone, Fell Sandstone and Scremerston Coal groups. Though the Fell Sandstone Group has always been recognized as diachronous, its base corresponds approximately with the top of the Tournaisian. Around Peel Fell this line is some distance above the top of the Lower Border Group, while in upper Redesdale and Coquet-dale to the east, it is below the top of the Main Algal Beds.

Around the head of Redesdale, Carboniferous strata rest directly on Lower Old Red Sandstone and Silurian rocks (Fig. 16). At the base there is generally a thin conglomerate, but at two places there are small outcrops of thicker, red conglomerates which have sometimes been classed as Upper Old Red Sandstone. These are at the head of Ramshope Valley, where there is a 15-m thick conglomerate with pebbles and boulders up to $\frac{1}{2}$ m across of Silurian shale and greywacke and Cheviot andesite; and on Windy Gyle where a conglomerate, at least 100 m thick, contains pebbles of Cheviot granite and andesite.

Apart from these conglomerates, the lowest Carboniferous strata form the Lower Freestone Beds which consist of flaggy sandstones, either red or grey and spotted with brown ochre, and purple, red, lilac and green shales with ochreous concretions. The thickness varies between 60 and 100 m, partly because the rocks rest on an uneven floor.

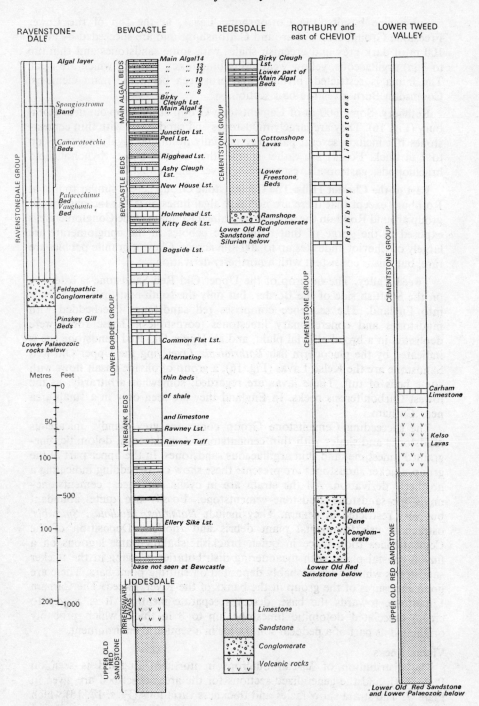

FIG. 16. *Comparative sections of Tournaisian and Upper Old Red Sandstone strata*

Submarine basalts, the Cottonshope Lavas, at the top of the Lower Freestone Group, crop out around Cottonshope and are succeeded by about 100 m of dark grey and greenish shales with some sandstones and thin (up to 3 m) argillaceous, yellow weathering limestones, some of which are algal. The fauna is restricted, bivalves and crustacea being the main elements. Coomsdon Burn gives the best section.

Rothbury. Some 600 m of Cementstone Group strata crop out near Rothbury (Fig. 16). They are largely mudstones and sandstones with thin cementstones but include several pale grey partially dolomitic algal limestones up to 7 m thick. Fossils are rather sparse; they include algae, rhynchonelloid brachiopods, gastropods, estuarine bivalves and ostracods.

East of the Cheviot Hills. The Cementstone Group here is similar to that at Rothbury except that there are no thick algal limestones. At the base of the group around Roddam the red coarse-grained Roddam Dene Conglomerate is exposed in the gorge of that name. The pebbles in the conglomerate are largely of Cheviot andesites up to 200 mm across. Cheviot granite pebbles are rare, but this is consistent with a northerly derivation.

Tweed Valley. The outcrop of the Upper Old Red Sandstone is extensive on the Scottish side of the Border, but only the south-east tip of it extends into England. The sequence comprises red sandstones interbedded with mudstones and concretionary limestones (cornstones). These rocks were deposited in a broad alluvial plain, and an Upper Old Red Sandstone age is indicated by the placoderm fish *Bothriolepis*. Overlying the Upper Old Red Sandstone are the Kelso Lavas (Fig. 16), a group of olivine-basalt flows with some beds of tuff. These lavas are regarded, somewhat arbitrarily, as the lowest Carboniferous rocks. In England they are seen only in a small area near Carham.

The succeeding Cementstone Group comprises grey, sandy, micaceous mudstones and shales with thin cementstones (argillaceous, dolomitic, fine-grained limestone) and thin argillaceous sandstones. In the upper part of the group, thicker sandstones are present: these show cross-bedding indicating a northerly derivation. All the strata are in cyclic sequences: cementstone–mudstone–sandstone–mudstone–cementstone. Fossils are quite abundant but very restricted in genera. They include *Naiadites*, *Modiolus*, *Spirorbis*, ostracods, fish fragments, plant debris and rootlets. Deposition of the Cementstones took place in quiet, brackish, shallow-water lagoons on a fluvial coastal plain, with meandering distributaries bringing in the thicker sandstones which were probably deposited on shifting point-bars. There are good exposures of the group in the banks of the River Tweed. The Carham Limestone, towards the base, deserves separate mention. It is a grey to cream brecciated dolomitic limestone, up to 8 m thick, which probably originated as part of a pedocal soil profile in a semi-arid environment.

Viséan Rocks

The distribution of Viséan outcrops in northern England is seen on Plate XIII, and the generalized sections for the areas described are given in Plate V. The strata show facies and thickness variations (Figs. 17, 18) which are related to the geographical setting shown in Fig. 13 that has been alluded to already.

Southern Isle of Man. The Basement Conglomerate, about 20 m thick and
resting on Manx Slates, is well exposed on the south-west shore of Langness,
where it is red-coloured, with pebbles of Manx Slates up to 40 cm across,
set in a gritty matrix. It passes up, by intercalations of hard calcareous grit
and limestone with small pebbles, into the overlying Castletown Limestones.
The conglomerate is a transgressive shoreline deposit, becoming younger as
it is traced north-eastwards.

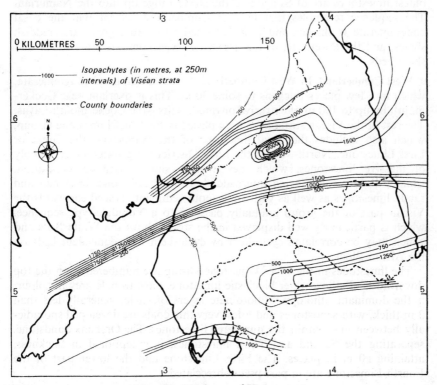

Fig. 17. *Thickness of Viséan strata*

The Castletown Limestones, which range in age from C_2 to the top of D_1,
comprise 100 m or more of grey and dark grey, relatively thin-bedded,
bioclastic limestones with thin shale partings and a prolific coral-brachiopod
fauna; gastropods and cephalopods are also common at some levels. Inland
exposures are few but the beds are well exposed on the foreshore in Castletown Bay and Derbyhaven. The overlying Poyllvaaish Limestones, 60 m
thick, are pale coloured and largely of reef-facies with a fauna which is very
rich, particularly in brachiopods and goniatites. They are well exposed only
on the shore at Poyll Vaaish. The highest limestones are the Black Limestones
or *Posidonomya* Beds, only 6 m thick, with a restricted fauna including
Goniatites falcatus Roemer.

Overlying the limestones is a group of submarine tuffs with some thin
tuffaceous limestones, agglomerates and basalt lavas—the Scarlet Volcanic

Group—which appears to be thrust over the limestones. As the whole Carboniferous outlier shows but gentle folding, the disturbances in the volcanic rocks may perhaps be due to progressive growth of a volcanic neck not far from the outcrop.

Northern Isle of Man. Beneath a cover of 45 to 80 m of glacial deposits in the northernmost part of the Isle of Man, boreholes have proved a limestone group about 250 m thick overlying 30 m of basal conglomerate. The oldest limestones are of S_2 age and the strata range up into the Namurian. The sequence resembles that in west Cumberland, except that the basal conglomerate is thicker and is interbedded with grey, green and reddish shales, and the lower part of the limestone series contains more sandy and shaly beds.

West Cumberland. In West Cumberland there is a thin basal conglomerate, which in a few places thickens to some 30 m. This is overlain near Cockermouth by up to 100 m of extrusive olivine-basalts—the Cockermouth Lavas. The rest of the Viséan strata have been placed in the Chief Limestone Group, which also includes the basal limestone of the Namurian, the Great or First Limestone. North-west of the Lake District the group consists mainly of massive limestones, which nevertheless show rhythmic deposition, 'standard' limestones being interbedded with porcellanous, bryozoan and sandy limestones as well as with terrigenous sediment. Traced eastwards the Viséan part of the group gradually passes into a Yoredale-type sequence, which is particularly well displayed in the strata above the White Beds and which may be correlated cyclothem by cyclothem with equivalent beds on the Alston Block.

The limestones in the Chief Limestone Group are numbered from the top downwards. Shallow-water bioclastic limestone, dark to pale grey in colour, is the dominant lithology. Subordinate layers of shale, generally less than 3 m thick, with sandstones and a few very thin coals are developed sporadically between and within the numbered limestones. The Orebank Sandstone, separating the Second and Third limestones, is exceptional in thickness, attaining 60 m in places. The Fifth Limestone and the lower part of the Fourth Limestone are in part pseudobrecciated.

Furness. In Furness, the Viséan succession is mainly calcareous. The Basement Beds, which range from 90 to 240 m in thickness, are shales and sandstones with conglomerates and thin limestones. They are poorly exposed and contain few fossils. Recently examined plant spores indicate a probable Tournaisian age for part at least of the Basement Beds. The Martin Limestone consists of 45 to 140 m of grey limestones, very fine-grained in places, with thin shale partings and contains the coral *Thysanophyllum pseudovermiculare* (McCoy). It is overlain by the Red Hill Oolite—a light grey oolitic fragmental, generally massive limestone, 50 to 60 m thick, of C_2 age. The Dalton Beds, with rich coral-brachiopod faunas, equate with the *Michelinia grandis* Beds of Garwood. They are from 115 to 260 m thick, thinning north-eastwards, and are grey and dark grey, relatively thin-bedded limestones, with shale partings particularly numerous about the middle. The Park Limestone of S_2 age is massive, white to cream coloured, without shale partings, and ranges from 125 to 140 m thick. At the top of the limestone succession comes the Urswick Limestone, 115 to 125 m thick, which is a pale, well-bedded limestone with thin shale partings; some beds are oolitic, and some pseudobrecciated. The highest Viséan strata, the Gleaston Group, consist of shales, thin sandstones and thin dark cherty or crinoidal limestones, with the '*Girvanella*' Band near the base.

Alston Block and north-east side of Lake District (Shap–Appleby). The lowest Orton Group rocks exposed near Shap are limestones, generally dolomitic, with the coral *Thysanophyllum pseudovermiculare* common at several horizons. The overlying and easily recognized Brownber Pebble Bed—a sandy oolitic limestone with white quartz pebbles—is followed by the *Michelinia grandis* Beds, which are thin, sandy and oolitic near Shap, thickening south-eastwards into massive limestones. The succeeding Ashfell Sandstone forms a series of fine-grained quartzitic sandstones with interbedded limestones. It is somewhat diachronous, extending over a greater time-span in the north than the south. Grey, white-weathering limestones of the overlying *Davidsonina carbonaria* and *Lithostrotion [Nematophyllum] minus* beds, are capped by the Bryozoa Band which includes about 10 m of variously shaly, porcellanous and fragmental limestones, with a rich bryozoan and brachiopod fauna.

Overlying Alston Group strata have at their base the thick-bedded bioclastic limestones, pseudobrecciated at some horizons, which carry the lowest *Dibunophyllum* Zone fossils, notably *Lithostrotion junceum* (Fleming): these form the Knipe Scar Limestone, about 60 m thick, broadly equivalent to the Melmerby Scar Limestone. The succeeding 250 m of rhythmic limestones, mudstones and sandstones form a Yoredale sequence almost identical with that of the Alston Block.

On the Alston Block, the Orton Group consists of a variable sequence of shales with limestones and thin sandstones, between 20 and 40 m thick, lying below the Melmerby Scar Limestone. These beds are known along the escarpment north of the Swindale Beck Fault, around the Cronkley Inlier and from the Roddymoor Borehole, near Crook, County Durham. South of the Swindale Beck Fault the Orton Group is represented by 90 to 180 m of strata comprising in upward succession the Ravenstonedale Limestones, the Ashfell Sandstone and the Hillbeck Limestones. The Ravenstonedale Limestones are impure limestones, sandstones and shales and include the *Thysanophyllum pseudovermiculare* Band and the Brownber Pebble Bed. The Ashfell Sandstone is a fine-grained quartz-sandstone and the Hillbeck Limestones are limestones and shales with a few thin sandstones.

At the base of the Melmerby Scar Limestone there is an erosion surface, marking a break in sedimentation. The limestone itself, with a thickness of some 35 m, is the thickest in the sequence. It is a pale grey, thick-bedded, bioclastic limestone with pseudobrecciation at various levels, especially towards the top, and contains a rich D_1 Zone fauna of corals and brachiopods. Above the Melmerby Scar limestone the relatively thin Robinson, Birkdale, Peghorn and Smiddy limestones are separated by shales, sandstones and seatearths. The Peghorn Limestone has two distinctive features: there is a thin band of algal nodules (the '*Girvanella*' Band) near the top, and the central part of the limestone is pale-coloured, massive and fossili-

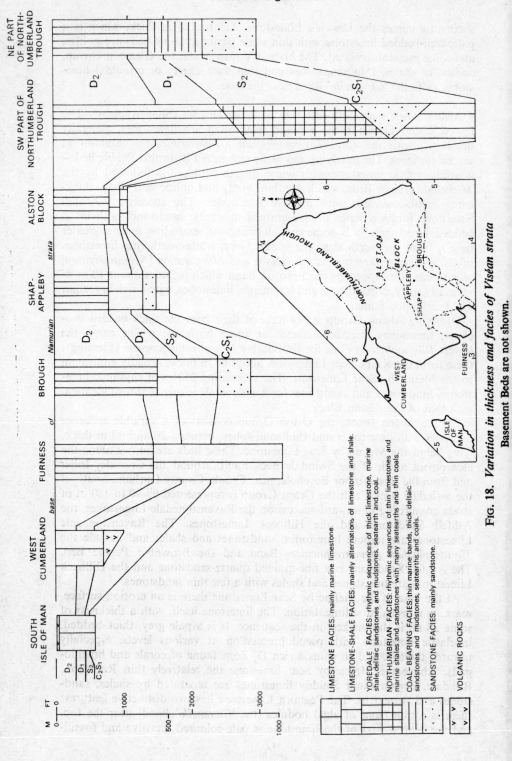

FIG. 18. *Variation in thickness and facies of Viséan strata Basement Beds are not shown.*

LIMESTONE FACIES: mainly marine limestone.

LIMESTONE-SHALE FACIES: mainly alternations of limestone and shale.

YOREDALE FACIES: rhythmic sequences of thick limestone, marine shale, deltaic sandstones and mudstones, seatearth and coal.

NORTHUMBRIAN FACIES: rhythmic sequences of thin limestones and marine shales and sandstones with many seatearths and thin coals.

COAL-BEARING FACIES: thin marine bands, thick deltaic sandstones and mudstones, seatearths and coals.

SANDSTONE FACIES: mainly sandstone.

VOLCANIC ROCKS

(*Cambridge University*)

A. Great Scar Limestone escarpment north of Brough under Stainmore; looking north-west

Plate VI

B. Syncline in Middle Limestone Group: Green's Haven, Berwick upon Tweed

(A 3042)

ferous with pale algal nodules near the base. The '*Girvanella*' Band is taken as the base of the D₂ Zone.

Higher strata, up to the Great Limestone at the base of the Millstone Grit Series, display the Yoredale type of cyclic sedimentation to perfection. Its characters have been noted above. Right across the Alston Block and the north-east side of the Lake District individual cycles may be traced with confidence, such is the regularity of their development. Individual limestones are characterized by particular lithologies or marker beds over certain areas; for example the Single Post Limestone is white and pseudobrecciated over a wide area, while the Scar Limestone generally includes some chert nodules and a band with *Lithostrotion junceum*.

Kirkby Stephen. When traced southwards across the Swindale Beck–Lunedale Fault, Viséan strata thicken abruptly, although all the cyclothems are represented both north and south of the fault. This area of relatively thick strata is developed over a narrow east–west zone between the Alston and Askrigg blocks in the Stainmore Trough (see Fig. 13). It does not extend west of about Tebay, but geophysical evidence suggests that it probably extends eastwards beyond Darlington.

North Cumberland and westernmost Northumberland. In the Northumberland Trough, Tournaisian strata are followed conformably by a rhythmic Viséan sequence which is thickest and most marine in the south-west.

The lowest Viséan strata are the Cambeck Beds, the uppermost division of the Lower Border Group. They comprise 200 m of rhythmically deposited sandstones, shales and thin shelly algal limestones. The fauna is dominated by bivalves and brachiopods, notably *Antiquatonia teres* (Muir-Wood), *Cleiothyridina glabristria* (Phillips), *Ovatia bioni* (Muir-Wood), *Schuchertella ambigua* (Muir-Wood) and *Syringothyris exoleta* North, but it also includes the Viséan coral *Palaeosmilia murchisoni* Milne Edwards and Haime. Algae are by no means so prominent as in the underlying Main Algal Beds, but are still quite common.

South-west of the watershed between the rivers North Tyne and Lyne the Middle Border Group (300–400 m thick) is lithologically similar to the Cambeck Beds, though individual sandstones thicken locally up to 75 m and seatearths with very thin coals appear. The sandstones are fine to medium grained, moderately well graded, and in many places show irregularly contorted bedding, probably caused by horizontal load-flowage under critical conditions of de-watering. Bivalves and brachiopods are the main fossils. The Whitberry Band at the base contains *Rugosochonetes cumbriensis* (Muir-Wood). In the Kershope Valley the 35-m thick Kershopefoot Basalt crops out towards the top of the Middle Border Group.

In northern Cumberland the Upper Border Group comprises some 600 m of rhythmic limestone–shale–sandstone–seatearth–coal sequences of S₂ age. The base of the Group is the Clattering Band or the equivalent Kingbridge Limestone, both of which mark the incoming of *Lithostrotion martini* Milne Edwards and Haime, while the top is the Naworth Bryozoa Band. Limestones in the group are generally detrital and less than 3 m thick (although a few reach 8 m). A few coals, up to 60 cm thick, have been worked locally. Near the base of the group is the only volcanic occurrence, the Oakshaw Tuff, up to 1 m thick, which lies in the general position of the Glencartholm Volcanic Beds of the Langholm district. In the southern part of their north Cumberland outcrop, strata now assigned to the Upper Border Group were formerly divided into the Birdoswald Limestone Group above and the Craighill Sandstone Group below. It is now clear that the latter is hardly less calcareous than the former and there is no justification for this lithological division.

In the Liddesdale Group, which succeeds the Upper Border Group, a rhythmic succession of Yoredale-type appears, with relatively pure bioclastic limestones generally between 4 and 15 m thick, as on the Alston Block; but in the Northumberland Trough the cyclothems are thicker because the sandstone and shale members are thicker. Correlation of the limestones is shown in Table 2. Sandstones are well sorted, with grain size from 0·1 to 0·5 mm, and are generally cross bedded. Liddesdale Group strata are well exposed in the Roman Wall area from Gilsland eastwards to the River North Tyne, where, with the intruded Whin Sill, they form a series of north-facing escarpments and southerly dip slopes. The Liddesdale Group contains typical *Dibunophyllum* Zone fossils, the most common being corals such as *Lithostrotion junceum*, *L. martini* and *Lonsdaleia floriformis* (Martin), brachiopods such as *Gigantoproductus maximus* (McCoy) and *Tornquistia polita* (McCoy), and bivalves. Goniatites are very rare.

West-central and north Northumberland. The Border and Liddesdale groups are defined in the western part of the Northumberland Trough, where the strata can be correlated on the basis of their marine fossils, but eastwards and northwards the beds become less fossiliferous and are divided on a lithological basis. The nomenclature is inconsistent, for not only are lithologies diachronous, but workers in different areas have taken dividing lines at different horizons. In general, the boundaries between groups are taken at higher horizons in north Northumberland than in western Northumberland.

The lowest Viséan strata, the Cambeck Beds and Middle Border Group, pass north-eastwards into the Fell Sandstone Group. With change of facies, the characteristic fossil horizons fail and it is not known to what extent the lithologically defined Fell Sandstone Group of Northumberland is diachronous. Around the headwaters of the North Tyne the group comprises about 450 m of fine-grained, generally greenish sandstones with subordinate unfossiliferous, red, purple, or greenish grey silty mudstones; there are a few seatearths. Irregularly contorted bedding is common in the sandstones. From Redesdale northwards through Rothbury Forest to Berwick upon Tweed, the group is between 240 and 300 m thick and has remarkably uniform lithology. Well-sorted, fine-grained quartzitic sandstones make up most of the group, the rest being thin pebbly sandstone bands and mudstone partings. Cross bedding and contorted bedding are common and indicate derivation from the north-east. Being almost wholly quartz-sandstones, the outcrop forms barren, heathery, moorland hills.

In west-central Northumberland strata approximately equivalent to the Upper Border Group form the Scremerston Coal Group, which is not defined palaeontologically. It is thicker than the equivalent strata in northern Cumberland, reaching a maximum of an estimated 2000 m around Falstone, between the Antonstown and Harrett's Linn faults. This increase is probably

TABLE 2

Limestone correlations

Askrigg Block	Alston Block	West Northumberland[1]	North Northumberland
Main	Great	Great	Great=Dryburn
Undersett	Four Fathom	Four Fathom	Sandbanks= Eight Yard
Three Yard	Three Yard	Three Yard	Acre=Six Yard
Five Yard	Five Yard	Five Yard	Eelwell=Beadnell =Nine Yard
	Scar	Scar=A	
Middle	Cockle Shell	Cockle Shell=B	
	Single Post	Single Post=C	? Budle
Simonstone	Tynebottom	Tynebottom=D	
Hardraw	Jew	Jew=E	Oxford
Gayle	Lower Little	Greengate Well=F	
Hawes	Smiddy	Middle Bankhouses	
Great Scar (top of)	Peghorn	Low Tipalt=Low Bankhouses	Upper Wishaw= Watchlaw
	Robinson	H	Middle Wishaw
		I	
		J	Lower Wishaw= Woodend
	Melmerby Scar (top of)	Denton Mill=K	Middle Penchford =Dun

(Middle Bankhouses, Low Tipalt=Low Bankhouses are bracketed together =G)

[1]The letters in the West Northumberland column indicate the correlation used on the original Geological Survey maps of this area.

correlated with contemporaneous movement along the faults, with extra subsidence and sedimentation between them (see Fig. 17). North of the Harrett's Linn Fault the thickness is about 1200 m in the Lewisburn–Plashetts area where the beds are well exposed in Lewis Burn. The strata show cyclic sequences similar to those in Cumberland, but with more terrigenous material and thinner, more argillaceous limestones. Coals and seatearths are thicker and more numerous; the thickest coal, which has been widely worked, is the Plashetts Coal, up to 2 m thick. The fauna is more restricted with fewer corals, but shows that the upper part of the group, above the Plashetts Dun Limestone, belongs to the Lower *Dibunophyllum* Zone. Sandstones are fine grained and quartzitic like those in the Fell Sandstone Group.

In North Tynedale and Redesdale, the top of the Scremerston Coal Group is taken at the base of the Redesdale Limestone. In Coquetdale and north-wards to Berwick the top of the group is the Dun Limestone, between 70 and 100 m higher than the Redesdale Limestone, which like the other limestones in this interval, appears to die out at about the Rede–Coquet water-divide. From here to the north of Alnwick the group is poorly exposed and only from 90 to 150 m thick, but still farther north it thickens again to about 275 m near Berwick where the upper strata are exposed in the sea-cliffs north of the town. Scremerston Group strata in north Northumberland are sandstones and shales with numerous seatearths and coals (nine or ten of which have been worked) and, especially in the lower part of the group, many thin argillaceous limestones with ostracods. Other fossils include *Lingula*, gastropods, bivalves and plant remains. The facies is deltaic, with a few marine incursions.

The upper part of the Viséan above the Denton Mill or Dun Limestone is lithologically similar to the Liddesdale Group described above, but shows some thinning as it is traced northwards. The limestones in general tend to thin and become split by shale partings. Coals, which are rarely more than a few centimetres thick in west Northumberland, tend to thicken northwards and are locally workable. Exposures are not good except for coast sections near Berwick upon Tweed and Sea Houses.

Millstone Grit Series (Namurian)

The regional palaeogeography of Millstone Grit times was in many respects similar to that shown as after the beginning of D_1 times on Fig. 13. The relatively greater subsidence in the 'troughs' as opposed to that over the 'blocks', coupled with an abundant supply of detritus from northern land masses, led to greater thicknesses of strata in some parts of the region than in others. Compare, for instance, the Woodland and Roddymoor sections on the Alston Block (Plate VII) with that at Throckley in the Northumberland Trough. In much of the area sedimentation continued throughout the period with only minor interruptions. There is strong evidence, however, that West Cumberland and the Vale of Eden, both close to the axis of the old Manx–Cumbrian ridge, received little or no sediment in the latter part of Millstone Grit times, and around Whitehaven sedimentation was minimal throughout the whole of the Namurian. These areas may even have stood out as land for part of the time.

For the most part, the environment in which the sediments of the Millstone Grit Series were deposited may be regarded as transitional from the marine–estuarine conditions of the Carboniferous Limestone Series to the deltaic lagoon–swamp conditions of the Coal Measures. The facies are essentially of two types. First, there is a 'Yoredale' facies consisting of repetitive sequences of limestones, marine shales and thin subordinate sandstones similar to those already described in the Carboniferous Limestone Series, although the

MODERN CLASSIFICATIONS		OLDER CLASSIFICATIONS	
		Northumberland, East Cumberland and Durham	West and Central Cumberland
WESTPHALIAN (Coal Measures)	*LOWER COAL MEASURES* Gastrioceras subcrenatum, Quarterburn or Swinstone Top Marine Band	*LOWER COAL GROUP* horizon of Ganister Clay Coal	*PRODUCTIVE COAL MEASURES* Harrington Four Foot Coal
NAMURIAN (Millstone Grit Series)	*MILLSTONE GRIT SERIES*	*'MILLSTONE GRIT'*	
		UPPER LIMESTONE GROUP OF *'CARBONIFEROUS LIMESTONE SERIES'*	*HENSINGHAM GROUP*
VISÉAN (Carboniferous Limestone Series)	Great Limestone	Great Limestone	Great (First or Main) Limestone

FIG. 19. *Table summarizing Millstone Grit Series stratigraphy in northern England*

limestones, with the exception of the Great Limestone, are generally much thinner. Second, there is a 'Millstone Grit' or more arenaceous facies, characterized by thick, coarse-grained cross-bedded sandstones, together with fine-grained sandstones, siltstones and mudstones, in which marine intercalations are comparatively thin. The entire sequence is of a rhythmic or cyclic character, and it is possible to trace some individual cyclothems for many kilometres where exposures are good. But correlation is not always easy, for cyclothems are sometimes incomplete, and the numerous impersistent 'channel' sandstones make direct comparison of sections difficult.

The occurrence of thin coal seams in both the 'Yoredale' and 'Millstone Grit' facies, together with sandstones that were apparently laid down in channels, shows that for much of the time the region was part of a wide deltaic area, with its surface at or close to sea level.

A major difficulty in the classification and correlation of the Millstone Grit Series of northern England is the dissimilarity between sequences of 'Yoredale' facies and the thick development of coarse gritstone and marine shales in the southern Pennines, where the traditional classification and zonation of the series has been based on goniatites. In northern England limestones and calcareous shales take the place of the dark goniatite-bearing mudstones and shales farther south; goniatites are rare and the majority of the remaining macro-fossils, though abundant at many horizons, are not particularly diagnostic of stratigraphical position. As a result, local classifications have been based primarily on lithology. These are summarized in Fig. 19 where they are related to the modern stratigraphical limits of the Series. The presence of limestones led earlier geologists to regard most of what we now know are Namurian rocks in the north of England as an upper division of the Carboniferous Limestone Series—the Upper Limestone Group or Hensingham Group. In recent years a diligent search for the rare goniatites, and a study of miospore assemblages, have resulted in a definition of the upper and lower limits of the Series, which as now defined is synonymous with the Namurian of north-west Europe. The base is taken at the first occurrence of *Cravenoceras leion* Bisat, though for practical reasons it has been generally agreed to regard the base of the Great Limestone which can be easily mapped, as the base of the Series. The top is taken at the base of a marine band containing *Gastrioceras subcrenatum* Frech in West Cumberland, and its apparent equivalents in Northumberland and Durham.

The following table shows the subdivision of the Series which holds good in the southern Pennines:

$$
\text{NAMURIAN} \left\{
\begin{array}{l}
\text{Yeadonian Stage } (G_1) \\
\text{Marsdenian Stage } (R_2) \\
\text{Kinderscoutian Stage } (R_1) \\
\text{Alportian Stage } (H_2) \\
\text{Chokierian Stage } (H_1) \\
\text{Arnsbergian Stage } (E_2) \\
\text{Pendleian Stage } (E_1)
\end{array}
\right.
$$

The stage symbols on the right of the table are the initial letters of the goniatite genera *Eumorphoceras*, *Homoceras*, *Reticuloceras* and *Gastrioceras*. In north-east England there is now sufficient evidence, much of it from deep boreholes, to say that all these major stages are present at least in part, though they are in general much thinner than their equivalents farther south. The disparity in overall thickness is further accentuated by the presence of non-sequences which probably represent pauses in deposition, one at least of them being of very long duration.

Plate VII gives detailed sections of Namurian successions throughout the Northern England region, the salient features of which are summarized below.

South Cumberland and Furness. In the area between the Duddon Estuary and Morecambe Bay to the south of the Manx–Cumbrian ridge, some 275 m of black shale with thin sandstones and a few limestones assigned to the upper part of the Gleaston Group (Dunham and Rose 1941) are thought to be Namurian in age. They are heavily drift covered and the limited evidence available suggests that the strata more closely resemble the basin-type facies of the Central Pennines south of the Craven Fault System.

Isle of Man. Namurian strata have been proved in bores on the Ayre along the northern coastline (Smith 1927). The succession may range up to about 260 m and consists of interbedded sandstones, shales and thin limestones, and compares more closely with the Namurian sequence of east Cumberland and the Alston Block rather than that of west Cumberland.

West Cumberland. The Namurian sequence begins with the First (Main or Great) Limestone, which ranges generally from 10 to 20 m in thickness. The Hensingham Group embraces all the remaining Namurian strata; at its base is a bed of gritstone, the Hensingham Grit, that is present for many kilometres along a discontinuous outcrop from Hensingham near Whitehaven to the vicinity of Cockermouth, where it fails. The rest of the group is variable, containing sandstones which are in places coarse, marine shales, sporadic thin and muddy limestones and a few thin coals.

The pause in sedimentation in the upper Namurian, mentioned above, apparently lasted throughout Chokierian, Alportian, Kinderscoutian and Marsdenian times. Yeadonian (G_1) marine shales containing the goniatite *Gastrioceras cumbriense* Bisat lie but a few metres above beds with a varied shelly assemblage of Arnsbergian (E_2) age which includes *Tylonautilus nodiferus* (Armstrong), though there is no sign of contemporaneous erosion in the strata, in the form of breccia, erosion surfaces, or even any discernible discordance of dip.

The Hensingham Group as a whole thickens markedly when traced northwards towards the Northumberland Trough. At its type area near Whitehaven it is about 50 m thick, at Workington it measures some 133 m, and north of Maryport more than 500 m have been proved in bores drilled through newer rocks. Part of this increase in total thickness represents a thickening of the individual beds, and part may be due to the presence of strata representative of some of the stages lacking farther south. Within the area of outcrop, however, present evidence favours placing most of the Group in the Arnsbergian (E_2) stage, and there may be non-sequence at the base as well as the top.

West of the Lake District the uppermost beds of the Hensingham Group are firmly placed in the Yeadonian (G_1) Stage by the presence of *Gastrioceras cumbriense* in its type locality. About ten metres higher in the sequence lie shales with *Gastrioceras subcrenatum* Frech, which mark the base of the Coal Measures. The intervening shales therefore represent beds which in the southern Pennines include the thick Rough Rock cyclothem.

In that part of the outcrop to the north of the Lake District the top of the Hensingham Group and of the Millstone Grit Series is less easily fixed, for there the *Gastrioceras subcrenatum* Marine Band has not yet been found: there are signs that the major intra-Namurian non-sequence may have lasted somewhat longer here, extending upwards to cut out the lowest part of the Coal Measures.

Vale of Eden. Namurian strata crop out on the western side of the Vale of Eden north-west of Penrith and in a narrow zone between Penrith and Kirkby Stephen. Although poorly exposed, evidence from bores indicates that they range between 400 and 425 m in thickness. The succession which is almost entirely reddened may be compared with an expanded Alston Block sequence and includes persistent marine bands and limestones. No strata younger than uppermost E_2 age have been proved.

Stainmore. In the Upper Carboniferous outlier of Stainmore, east of Brough in Westmorland (Owens and Burgess 1965), Namurian strata consist of sandstones, shales and coals, totalling 457 m, and include up to 20 marine marker beds in addition to limestones. The presence of all the stages has been established with the exception of the *Homoceras* stages, which are only inferred.

Durham and Northumberland. Namurian strata crop out widely over and to the east of the Pennines, from Teesdale in the south to Tynedale in the north, and thence extending north-eastwards to reach the coast between Amble and Alnmouth. The Series ranges in thickness between about 240 and 533 m being thinnest on the Alston Block, and thickest in the Northumberland Trough. As Plate VII shows, the bulk of the thickness variation takes place within the Pendleian (E_1) Stage. The E_2, H, R and G_1 stages, though very thin in comparison with their southern Pennine equivalents, show more uniformity of thickness and accord with a progressive northward thinning across northern England. It appears that the relative movement of subjacent crustal blocks had by this time almost ceased in north-eastern England.

The series comprises a lower 'Yoredale'-type facies and an upper arenaceous facies which in terms of the former classification are equivalent respectively to the 'Upper Limestone Group' and the lower part of the Durham 'Millstone Grit'. The Viséan–Namurian boundary is approximately at the base of the Great Limestone, and for reasons which are admirably summarized by Johnson and others (1962) it is drawn at the base of that bed. In Durham the upper part of the sequence consists of a number of thick coarse sandstones or 'grits' separated by relatively thin argillaceous measures which include marine shales. The onset of this 'Millstone Grit' arenaceous facies was later than in the mid-Pennines. In Northumberland, sandstones are numerous throughout the sequence, although the 'grits' forming the top of the Durham sequence cannot be traced throughout the county.

Although faunal and microfloral evidence indicates that all the stages of the Namurian are represented, non-sequences and penecontemporaneous erosion surfaces are common in both E_1 and E_2 stages; and on the Alston Block evidence from the Woodland Borehole suggests that parts of the R_1 and R_2 stages, also, are not present, possibly due to non-deposition.

Coal Measures (Westphalian)

Many of the factors controlling sedimentation during the Namurian continued to operate through Coal Measures times. A great land mass persisted to the north and north-west beyond the Highland Boundary Fault in Scotland and sediment carried down by the rivers which drained it poured into the subsiding gulf which was bounded on the south by an island mass stretching from central Wales to Brabant. The total thickness of the resultant strata was controlled mainly by the amount of subsidence which took place; and nowhere in northern England does this approach the maximum for the basin as a whole which is found in the vicinity of Manchester. There are signs that 'humps' existed in the floor of the basin where subsidence was less than in surrounding areas; for instance, the Coal Measures in west Cumberland thin towards the Lake District massif as if this ancient submerged block were still exerting its influence upon sedimentation in Coal Measures times.

The cyclic nature of the deposits is as marked in the Coal Measures as in the underlying Carboniferous rocks, the marine phase being less important and the coal-forming phase more so. The complete cyclothem consists of a marine shale bed, overlain by non-marine shale or mudstone, then sandstone, a rootlet bed, or seatearth, and a coal seam, the latter representing a phase when sediments were built up to water level after each pulse of subsidence. Most cyclothems are incomplete and lack one or more of these elements. Individual cyclothems vary from 1 to 30 m or more, the thinnest commonly being mere alternations of rootlet beds and mudstone. This cyclic sequence continued throughout the Lower and Middle Coal Measures, which total about 450 m in Cumberland and 600 m in Northumberland and Durham.

Since sedimentation more or less kept pace with subsidence, the succession is mainly deltaic or estuarine in character, each marine band representing comparatively brief episodes of marine incursion which allowed the existence of such creatures as *Lingula*, goniatites, pectinoids and, less commonly, productoids. The seas during the marine episodes were too muddy to allow the deposition of thick limestone beds, though thin bands of calcareous shale occur in association with some marine bands. Mudstones with non-marine bivalves (mussels) and ostracods represent brackish conditions as the sea receded; and the increase in coarseness of sediment marks the approach of a delta front building the deposits up to water level once more. On the flat delta surface, at or near sea level, flourished extensive swamp forests of giant club mosses (Lycopods) and fern-like trees (Pteridosperms) that grew and died, building up layers of peaty debris. The forests died in the inundation which commenced the next cycle, and compaction and the long series of chemical and physical changes that have converted the peaty accumulations into coal began at once.

In places the channels of rivers flowing across the delta swamps cut down through the unconsolidated strata and were then filled by sand and silt.

These are the common cause of 'washouts', 'wants' or 'nips' in coal seams where the coal ends against barren rock. In the Cumberland coalfield several square kilometres of the Main Band coal are absent in a 'nip' near Workington. The Harvey Seam in Durham is likewise affected by numerous 'washouts' and a dendritic pattern of old river courses can be made out on a plan of the seam.

Because of the concentration of plant and animal debris, and because the water table was hardly ever far below the surface, the newly formed sediment was in a state of chemical reduction. Vast amounts of iron derived from the weathering of the northern land mass were imprisoned in the strata, mostly in the ferrous state, the bulk of it in the form of siderite, with a minor amount of sulphide. The colour of the rocks, which largely depends on the included iron compounds, is therefore mainly grey in the shales and grey to yellowish brown in the sandstones. Variations of shade in the range grey to black are a reflection of the amount of organically-derived carbon present.

Soon after the beginning of Upper Coal Measures times a change took place in Cumberland; the details are not yet fully understood, but the result was that the contained iron was now deposited as red oxide, which was not reduced but remained as a red pigment in the rock. Some think that the red colour is a reflection of the nearness of the weathering land mass providing the sediment. Again, although some beds resembling seatearths are encountered in the red beds sequence there may not have been enough organic carbonaceous matter to maintain a reducing state, for animal remains are also rare.

Apart from the Coal Measures that were red at the time of deposition, there exist rocks in which the iron minerals have clearly undergone a secondary process of oxidation after deposition as grey measures. These are found close below the unconformity at the base of the overlying Permo-Triassic rocks; and the zone of reddening transgresses the strata in concordance with the unconformity. Criteria exist for distinguishing secondary from primary red beds, but these are not always definitive and confusion between the two types in Cumberland has led to much obscurity in stratigraphy. For instance, the red Coal Measures of the St. Bees and Whitehaven area which include the Whitehaven Sandstone of Whitehaven are now considered to be Middle Coal Measures in which the iron compounds have been oxidized below a late Carboniferous or early Permian land surface.

Formerly these beds were placed in a 'Whitehaven Sandstone Series' and equated generally with the Upper Coal Measures.

In the Lower and Middle Coal Measures the dominant rock types are shale or mudstone, and sandstone. The colour varies from dark to light grey, the marine shales usually being dark. Sandstones are generally medium to fine grained though in both the Cumberland and Durham coalfields there are thick beds of coarse sandstone in the Lower Coal Measures which have been mistakenly correlated with the Millstone Grit of other areas. Seatearths, leached by the action of plant roots, vary from light-coloured fireclays to mudstone with rootlets, and the more arenaceous seatearths from sandstone to hard siliceous ganister. Intermediate types of rock, interlaminated sandy mudstone or shaly sandstone, occupy considerable thicknesses of strata.

Classification

Fig. 20 shows how the Coal Measures of northern England are sub-divided on the basis of plants, plant spores, and mussels. The marine strata in a few places contain a rich fauna of goniatites, brachiopods and bivalves, though more commonly they carry only *Lingula* and microfossils. Since the marine fossils are confined in thin widespread bands, their main value is in providing marker horizons which help to give precise correlations not only within individual coalfields but, in many instances, throughout the rest of Britain and north-west Europe. On the continent subdivision of the West-phalian has long been based on plants, but in the last 40 years or so in this

MAJOR DIVISIONS	PLANT CLASSIFICATIONS		NON-MARINE BIVALVE ZONES	IMPORTANT MARINE BANDS
Stubblefield and Trotter 1957	Heerlen Congress 1927	Miospore Zones		
UPPER COAL MEASURES	Westphalian D	*Thymospora obscura*	*Anthraconaia prolifera* and *Anthraconauta tenuis*	
—M——M—	Westphalian C	*Torispora securis*	*Anthraconauta phillipsii*	*Anthracoceras cambriense* (St. Helens, Down Hill)
		—M——M—	—M——M—	
		Vestispora magna	Upper *Anthracosia similis* and *Anthraconaia pulchra*	*Anthracoceras hindi*
MIDDLE COAL MEASURES	—M——M—		—M——M—	*A. aegiranum* (Bolton, Ryhope)
	Westphalian B		Lower *A. similis* and *A. pulchra*	
		Dictyotriletes bireticulatus	Upper *Anthraconaia modiolaris*	
—M——M—	—M——M—		—M——M—	*Anthracoceras vanderbecki* (Solway, Harvey)
		Schulzospora rara	Lower *A. modiolaris*	
LOWER COAL MEASURES	Westphalian A	*Radiizonates aligerens*	*Carbonicola communis*	
		Densosporites anulatus	'Anthraconaia' lenisulcata	*Gastrioceras subcrenatum* (Quarterburn)
—M——M—	—M——M—		—M——M—	
MILLSTONE GRIT SERIES	Namurian			

(From Edwards and Trotter 1954, p.46, with additions and amendments)
—M—M— *marine band*

FIG. 20. *Table summarizing Coal Measures stratigraphy in northern England*

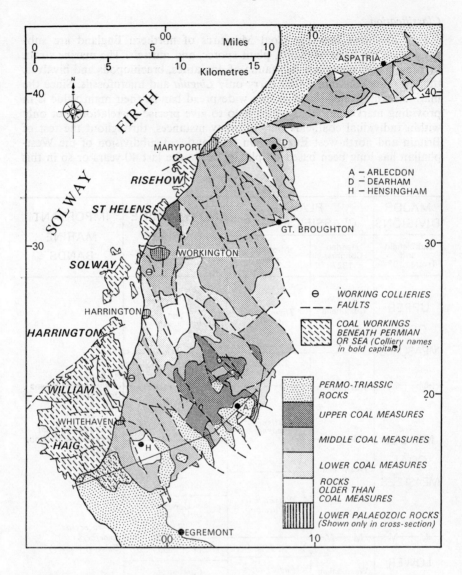

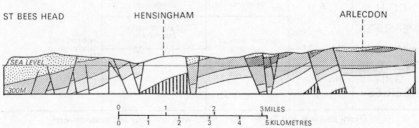

Fig. 21. *Map and section of the Cumberland Coalfield*

country more reliance has been placed on the mussels taken in conjunction with the marine bands.

Of the mussel genera, *Carbonicola* was dominant in the Lower Coal Measures but it did not survive the Solway–Harvey marine incursion and was replaced in the Middle Coal Measures by *Anthracosia*. *Naiadites* was common throughout the Lower and Middle Coal Measures, but disappeared near the time of the St. Helen's–Down Hill Marine Band and was replaced by *Anthraconauta* which was dominant in the *phillipsii* and *tenuis* zones of the Upper Coal Measures. Other fossils commonly preserved in the mudstone of the Coal Measures include the ostracod *Geisina* which occurs abundantly in the Lower Coal Measures and again in the Upper *similis–pulchra* Zone, and '*Estheria*' which tends to occur in isolated bands or in association with marine forms.

There is little doubt that the Coal Measures of northern England were laid down as one continuous thick sheet of sediment, vast quantities of which were removed by erosion following the late-Carboniferous earth movements. Erosion has been greatest in the tectonically 'positive' areas of the Lake District and the northern Pennines with the result that here Coal Measures are preserved only in small patches on the downthrow sides of major faults. The larger coalfields of Cumberland and Durham and Northumberland are in areas where there has been no long-continued history of isostatic elevation.

West Cumberland. Here the heavily faulted strata dip towards the Irish Sea and the Solway Firth in radial fashion away from the Lake District massif (Fig. 21). The succession, well known from sections obtained in bores and mines along the West Cumberland coastal strip, is condensed; at its thickest—450 m in the Workington area—it is less than one-third of that in the Manchester Coalfield, but all the mussel zones are present, as are at least ten of the known marine bands (Fig. 23). The strata below the Upper Three Quarters seam show great lateral constancy over wide areas in the coalfield, though some of the sandstones vary greatly in thickness. Many of the individual beds can be traced for several kilometres in borehole and shaft sections along the coastal belt and out to sea. A similar constancy is a feature of beds of the same age in the coalfields flanking the south Pennines. But this uniformity is not maintained to the north-east along the south side of the Solway Plain where the marine bands below the Albrighton fail, and the major non-sequence known elsewhere in Cumberland to cut out the Namurian *Homoceras* and *Reticuloceras* zones and part of the *Gastrioceras* (G$_1$) Zone of the Pennine sequence also cuts out the *A. lenisulcata* and *C. communis* zones. Above the Upper Three Quarters seam the beds are much more variable both in thickness and in lithology. Locally thick sandstones thin out laterally or pass into mudstones. There is much lateral splitting of thicker coals into separate seams. For instance, the Main Band of Whitehaven splits into the Cannel, Metal and Crow coals north of the River Derwent, and is absent altogether in a 'washout' near Workington. The Bannock Band of Whitehaven is represented by two separate seams in other parts of the coalfield.

The Upper Coal Measures are not nearly so well known as the beds below, for they carry no workable coals. In the coastal belt they have been encountered in bores north of Workington and in offshore workings south-west of Maryport. Near Workington the lowest few metres immediately above the St. Helen's Marine Band are normal grey shales and sandstones with seatearths and at least one thin coal, but it is plain from boreholes near Frizington, Aspatria and Crookdake that the bulk of the succession is of 'red beds'—red and purple shales, mudstones and sandstones, with a few rootlet-beds and at least two thin fresh-water limestones, which contain the small coiled shell of the annelid *Spirorbis*.

The greatest known development of Upper Coal Measures in the region lies beneath the Permo-Triassic rocks of the Solway Plain where they have been proved in bores near Aspatria. Here the fossil assemblages in the shales include *Anthraconaia pruvosti* (Chernyshev), *Anthraconauta phillipsii* (Williamson) and *A. tenuis* (Davies and Trueman) with *Euestheria simoni* (Pruvost) and *Leaia bristolensis* Raymond, and these are diagnostic of the *A. tenuis* Zone. An approximate estimate of 300 m of concealed Upper Coal Measures along the south side of the Solway Plain is based almost entirely on the red colour of the rocks as proved in boreholes. There are signs, however, that a 'red beds' facies in lower strata develops eastwards towards the Eden Valley where a bore at Barrock Park between Wreay and Southwaite has proved a red mudstone with mussels of the *A. modiolaris* Zone.

Northumberland and Durham. The western edge of the Lower Coal Measures outcrop is a denticulate line extending from Amble in the north to the vicinity of Barnard Castle in the south (Fig. 22). The measures dip generally eastwards to the coast, and continue under the North Sea, where they have been proved in offshore borings by the National Coal Board to a distance of 7·5 km offshore. The south-eastern part of the coalfield is con-cealed beneath Permian rocks, but since the eastward inclination of the basal Permian unconformity is little more than the general fall of the ground towards the coast the thickness of the Permian rocks has not inhibited exploration and working of the concealed Coal Measures, which are almost as well known here as in the exposed coalfield.

As in Cumberland, the sequence is a cyclic one of coal, shale and sandstone (Fig. 23). Marine bands occur and have been correlated with similar bands in other coalfields. In the lower part of the Lower Coal Measures the existence of a thick bed of gritstone—the 'Third Grit' of Durham—led early workers to equate this part of the succession with the Millstone Grit of the Pennine area. The discovery below the Third Grit of the Quarterburn Marine Band, which is inferred to be the equivalent of the *Gastrioceras subcrenatum* Marine Band, enables a true base to be fixed for the Coal Measures.

The Lower and the Middle Coal Measures total some 725 m, and most of the productive coals are concentrated in the middle of these strata from the Brockwell to the High Main. Below the Brockwell seam, sandstones tend to be coarse and some are siliceous enough to be termed ganister—the Tow Law Ganister is an example. The Middle Coal Measures above the High Main include massive sandstones, examples of which are the High Main Post, the Seventy Fathom Post and the Grindstone Post—the latter having been worked for the well-known Newcastle Grindstones.

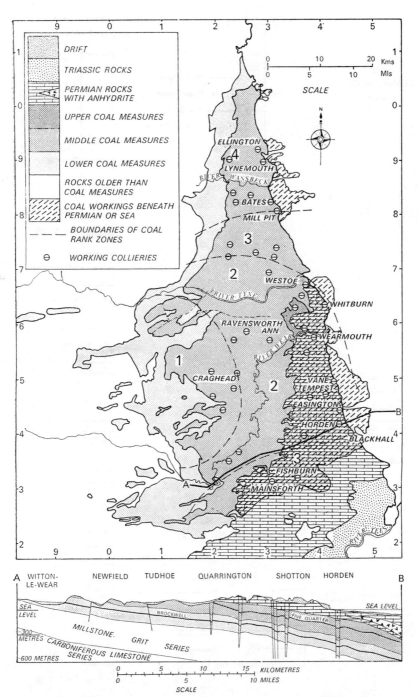

FIG. 22. *Map and section of the Northumberland and Durham Coalfield*
Zones are numbered 1 to 4 in decreasing order of coal rank.

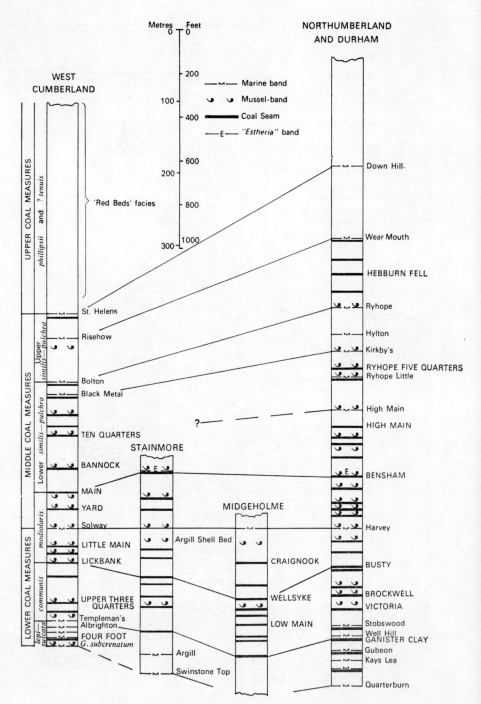

FIG. 23. *Correlation of the Coal Measures in northern England, showing marine bands and division into non-marine bivalve zones*

Upper Coal Measures are present in only two small areas in the coalfield. Near Boldon, in Durham, the Down Hill Bore proved mudstones and siltstones with subordinate sandstones and a few thin coals near the axis of the Boldon Syncline. Fossils at several horizons include *Anthraconauta phillipsii* and '*Estheria*'. Below lies the Down Hill Marine Band, which is equated with the Top Marine Band of the East Midlands, where it marks the base of the Upper Coal Measures. This information, supplemented by a section containing the marine band in the banks of the Wear at Claxheugh, and the occurrence of *A. phillipsii* in shales from a temporary exposure at Hylton Castle, leads to the belief that about 150 m of Upper Coal Measures may be preserved west of Sunderland. Near Killingworth in Northumberland it is inferred entirely on structural grounds that some 155 m of Upper Coal Measures probably exist on the northern (downthrow) side of the Ninety Fathom Fault. By contrast with Cumberland the Upper Coal Measures in Northumberland and Durham are of grey facies.

Midgeholme, Stainmore and Canonbie. There are three other small coal-fields in northern England. One is the Midgeholme Coalfield in the northern Pennines, where some 210 m of strata including the Solway Marine Band are preserved on the north side of the Stublick Fault in three small patches. Another lies south-east of Brough, where the Swinstone and Argill Marine Bands are equated with the Quarterburn and Kays Lea marine bands of Durham, and an '*Estheria*' Band near the top of the sequence gives correlation with that above the Bensham Seam of Durham. The third is in the extreme southern part of the Canonbie Coalfield, which lies mainly in Scotland. Near Riddings, reddened Upper Coal Measures are presumed to overlie productive coal measures.

6. Igneous Rocks associated with Carboniferous Strata

Both volcanic and intrusive rocks are associated with the Carboniferous strata of northern England. The volcanic rocks have been touched on in the account of the stratigraphy; they are confined to the Lower Carboniferous. Mainly olivine-basalts, they include the Kelso Lavas beneath the Cement-stones in the Cheviot region (Fig. 16), lavas and tuffs near the western end of the Border between Bewcastle and Kershopefoot, the Cockermouth Lavas—a thick series of olivine-basalts between the basal conglomerate and the Seventh Limestone of Cumberland—and the Scarlet Volcanic Group in the Isle of Man. The last named is the highest group of Carboniferous volcanic rocks in the region, and consists of porphyritic olivine-basalts and basaltic tuffs and agglomerates. The group apparently overlies the Black Limestones above the Poyllvaaish Limestones, but its actual relationship to the adjacent rocks is far from clear.

The intrusive rocks comprise sills and dykes. Some of the latter are of Tertiary date and will be dealt with at a later stage (p. 82). Those now considered comprise the Whin Sill and its associated dykes, which form a single petrographic province. It has been suggested that the dykes represent the feeding conduits through which the magma flowed before solidification. Both the sill and the dykes are mainly fine- to medium-grained quartz-dolerite, dark in colour. Small phenocrysts of plagioclase feldspar occur sparingly, especially in the dykes that rise into the Coal Measures, and in the Little Whin Sill. Coarser pegmatitic varieties occur locally in the Whin Sill, and along the margins of the intrusions the rock has in places been altered by chemical replacement to a form known as white whin. The rock from both sill and dykes is exceedingly tough and makes excellent roadstone.

By far the most important intrusion is the great Whin Sill itself, which crops out at intervals right across Northumberland and north-east Cumberland, along the Pennine escarpment and in Teesdale. It has also been proved in many boreholes between the western outcrops and the east coast, and its total area must be at least 5000 sq km. Over wide areas the sill maintains a general position within the Carboniferous Limestone Series, and it is not surprising that many geologists, including Phillips, regarded the mass as a contemporaneous lava flow, though its intrusive origin was advocated as early as 1826 by Sedgwick, and established by Tate about 1870. Abrupt changes in position occur, usually along joint and fault planes. In the Cronkley area of Teesdale it is in the Melmerby Scar Limestone, and at Kyloe in north Northumberland it is as low as the Fell Sandstones. On the other hand on the extreme northern edge of the Alston Block its horizon is much higher—at Midgeholme it is intruded into the Lower Coal Measures—and south of the Lunedale Fault near Selset its horizon is in the vicinity of the Four Fathom Limestone. A maximum thickness of about 73 m has been recorded at the Burtree Pasture lead mine in Weardale, but the average may lie between 25 and 30 m. At many localities the intrusion consists of two or more layers, which may be separated by some hundreds of metres. The uppermost leaf is in places referred to as the 'Little Whin Sill', a good exposure of which is seen in Greenfoot Quarry, where it is some 12 m thick. In drift-free areas the Whin Sill gives rise to bold escarpments such as may be seen in Teesdale, along or north of the Roman Wall, and again in the Belford country. It forms sea cliffs in the Farne Islands; Bamburgh and Dunstan-burgh castles are built upon it; and it is responsible for many waterfalls, including High Force (see Plate I, *Frontispiece*). Columnar jointing is well displayed by the Whin Sill in most of its main outcrops.

Stratigraphic and radiometric evidence suggests that the Whin Sill was intruded late in Carboniferous times, and an age of 295 ± 6 m.y. has been computed (Fitch and Miller 1967). The intrusion is of interest in that it was the subject of pioneer isotopic age determinations by Arthur Holmes.

Four major dyke-echelons are related to the Whin Sill, and all trend roughly east-north-east. They are, from north to south, the Holy Island, Lewisburn–High Green, St. Oswald's Chapel and Hett dyke-echelons, and the sill appears to be essentially confined between the most northerly and most southerly of these echelons. The Hett dyke provides the longest un-broken line of outcrop, for it can be traced for 32 km across the Coal Measures of south Durham. By contrast, the Lewisburn–High Green system, which stretches for at least 80 km from west of the North Tyne near the Border to the Northumberland coast south of Alnmouth, is broken into a large number of roughly east–west trending segments, offset along an east-north-east line.

7. Permian and Triassic Systems

In late Carboniferous and early Permian times, northern England was uplifted and deeply eroded following the Hercynian (Armorican) earth movements. West of the present Pennines, an immature erosional topography was gradually buried by thick accumulations of continental detritus and wind-blown sand, but Permian and later sediments to the east lie upon a mature eastwards-sloping peneplain. The sub-Permian surface was reddened by desert weathering to depths of up to 150 m in places in the north-west, but the depth of reddening rarely exceeds 100 m in the east and is more generally 6 to 12 m.

Sediments deposited upon the reddened surface are conventionally classed as Permian and Triassic, the junction between these being taken at the incoming of thick red water-lain sandstones above red mudstones and siltstones. This incoming is demonstrably diachronous, however, and can only broadly be related to internationally acceptable chronostratigraphic boundaries. For this reason it is convenient to discuss deposits of the two systems together. Permian and Triassic outcrops and representative sequences are shown in Figs. 24 and 25.

North-east England

Permian rocks crop out widely in coastal areas of County Durham, and extend south-westwards to Darlington and beyond. There are three small outliers on the Northumberland coast, and the Permian has also been proved in a number of boreholes offshore. Sediments traditionally assigned to the Trias are found mainly in a belt between Teesside and Darlington (Fig. 24). The Permian sequence in north-east Durham is shown in Fig. 26, and a rather fuller sequence for south-east Durham in Table 3.

The Lower Permian is represented in north-eastern Durham by up to 60 m of weakly cemented aeolian sandstone, the 'Yellow Sands', which form long ridges trending west-south-west, interpreted as seif-dunes, and in south-eastern Durham by thin discontinuous piedmont breccias. The sands were partly redistributed during the ensuing Zechstein Sea transgression, but the breccias were relatively stable and were redistributed only locally. The sands are generally similar to deposits of the same age which form the North Sea gas reservoirs.

The fish-bearing sapropelic Marl Slate (the German Kupferschiefer), the base of which is taken as the base of the Upper Permian, ranges up to 5·5 m in thickness but is generally 0·6 to 1·2 m thick and is locally absent over eminences in the pre-Permian basement. It passes uninterruptedly over the dune ridges, suggesting that the deepening of the Zechstein Sea following its transgression was extremely rapid, and is succeeded by up to 73 m of evenly bedded fine-grained dolomites of the Lower Magnesian Limestone. These generally yield a sparse fauna of bivalves and bryozoa, but contain a more varied fauna including productoids at a few localities (notably Raisby Hill

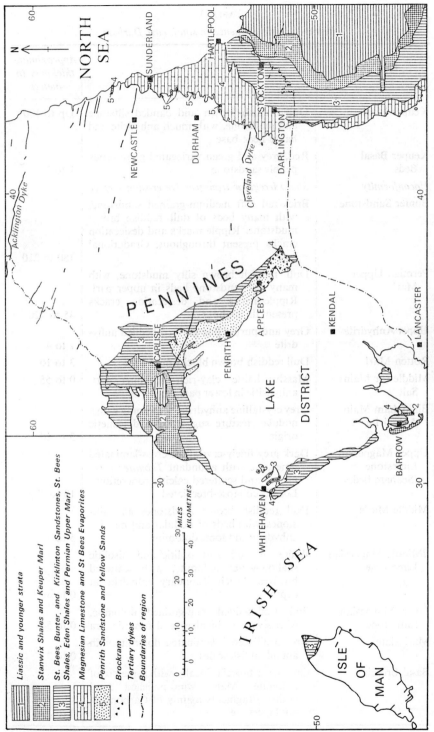

FIG. 24. *Distribution of Permian and Triassic rocks*

TABLE 3

Permo-Triassic Sequence in south-east Durham

Formation		Approximate thickness in metres
Keuper Marl	Dull reddish brown and banded siltstone and mudstone, with much anhydrite and halite near base	Up to 100
Keuper Basal Beds	Red, grey and green variegated gypsiferous pebbly sandstone	1 to 2
Unconformity	*Sharp irregular transgressive erosion surface*	
Bunter Sandstone	Brick-red soft medium-grained sandstone, with many beds of dull reddish brown mudstone. Ripple marks and desiccation cracks present throughout. Gradational base	180 to 210
Permian Upper Marl	Dull reddish brown silty mudstone, with many red sandstone beds in upper part. Ripple-marks and desiccation cracks present throughout	45 to 120
Upper Anhydrite	Grey and purple fine-grained bedded anhydrite	1 to 4
Rotten Marl	Dull reddish brown blocky mudstone	3 to 10
Middle (or Main) Salt	Massive halite, clay-rich in upper part, anhydritic in lower part	0 to 55
Billingham Main Anhydrite	Grey crystalline anhydrite, commonly with a nodular texture suggesting a diagenetic origin	3 to 10
Upper Magnesian Limestone (Seaham Beds)	Dark grey finely-crystalline cross-laminated limestone with abundant *Tubulites permianus* and scattered calcitic concretions. Locally collapse-brecciated	10 to 27
Middle Marls	Dull reddish brown mudstones and siltstones, with beds of nodular and massive anhydrite and some dolomite	5 to 37
?Middle Magnesian Limestone	White to buff soft oolitic and pisolitic shallow-water dolomite, with scattered bivalves. Much secondary anhydrite at depth	0 to 46
Lower Magnesian Limestone	Buff well-bedded fine-grained dolomite, with scattered bivalves and foraminifera	0 to 50
Marl Slate	Grey argillaceous laminated dolomite with abundant fish scales	0 to 3
Basal Breccia	Grey hard breccia, locally with a matrix of dolomite. Many wind-polished sand grains. Fragments mainly of Carboniferous limestone	0 to 10

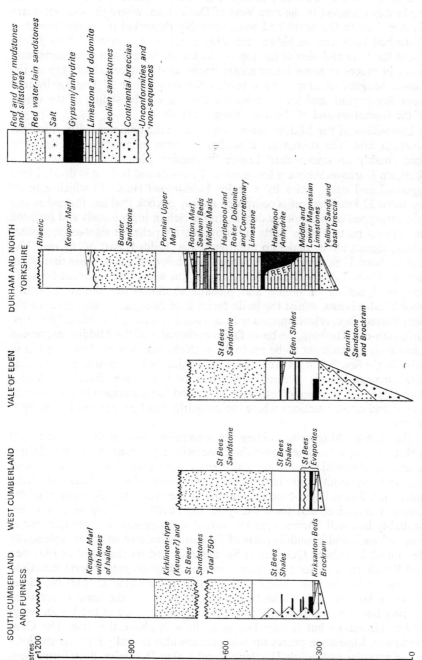

Key:
- Red and grey mudstones and siltstones
- Red water-lain sandstones
- Salt
- Gypsum/anhydrite
- Limestone and dolomite
- Aeolian sandstones
- Continental breccias
- Unconformities and non-sequences

SOUTH CUMBERLAND AND FURNESS
- Keuper Marl with lenses of halite
- Kirklinton-type (Keuper?) and St Bees Sandstones Total 750+
- St Bees Shales
- Kirksanton Beds
- Brockram

WEST CUMBERLAND
- St Bees Sandstone
- St Bees Shales
- St Bees Evaporites

VALE OF EDEN
- St Bees Sandstone
- Eden Shales
- Penrith Sandstone and Brockram

DURHAM AND NORTH YORKSHIRE
- Rhaetic
- Keuper Marl
- Bunter Sandstone
- Permian Upper Marl
- Rotton Marl
- Seaham Beds
- Middle Marls
- Hartlepool and Roker Dolomite and Concretionary Limestone
- Hartlepool Anhydrite
- Middle and Lower Magnesian Limestones
- Yellow Sands and basal breccia
- REEF

Metres scale: 1200, 900, 600, 300, 0

FIG. 25. Representative sections of Permian and Triassic rocks in northern England

and East Thickley quarries) in south Durham. Foraminifera and bivalves are locally common in the area west of Darlington, where the sediments are thinner than to the north and were probably deposited in shallower water. Disturbed beds and turbidites are widespread at a horizon 2·5 to 3·7 m above the base and also at the top of the Lower Magnesian Limestone, and occur in places at some intermediate levels. Disturbances at the top of the Lower Magnesian Limestone extended for more than 27 km northwards from Sunderland, and locally involved the removal by sliding of the whole of the formation and of the underlying Marl Slate.

Deposition of the Middle Magnesian Limestone commenced in northern Durham after the submarine disturbances were completed, but followed conformably on undisturbed Lower Magnesian Limestone in central and southern Durham. Above a few metres of transitional beds, it is divided into lagoonal and basin facies by a major barrier reef (Fig. 26) which extends for some 32 km. The reef is composed mainly of rock that has formed *in situ* and entrapped contemporary debris, and is rich in brachiopods and bryozoa in its lower parts. Higher parts are progressively richer in algal-stromatolites which, almost alone, make up the uppermost (bedded) part of the reef. The reef is about 37 m thick around Sunderland, but thickens to more than 90 m farther south. Locally it may be up to 1·6 km wide. The lagoonal beds are composed mainly of oolitic and pisolitic shallow-water dolomite with a poor bivalve fauna, whilst the basin facies is of fine-grained dolomite which, near the reef face, is interspersed with eastwards-thinning fans of reef detritus. Off south-east Durham the basin facies carbonates of the Middle Magnesian Limestone are thought to be up to 21 m thick, but they are less than 0·3 m thick in the South Shields area where only a thin bed of nodular or columnar algal-stromatolites is present. This bed underlies the thick Hartlepool Anhydrite in a number of offshore bores and the corresponding residue in inland and coastal sections where the anhydrite has been removed by natural solution.

The Upper Magnesian Limestone comprises four main divisions, all with a limited fauna. The lowest, the Concretionary Limestone, is essentially a laminated dolomitic limestone which underwent extensive collapse-brecciation and dedolomitization in its lower parts during solution of the underlying anhydrite. The collapse-breccias are one of the characteristic features of the Durham coastline, and form many of the headlands. They are, however, probably less well known than the varied and commonly grotesque calcite concretions found in middle parts of the formation, and which are splendidly developed in Fulwell Quarries at Sunderland, and on the coast at Marsden and Whitburn. The Concretionary Limestone displays great lateral variation, and passes locally into almost pure dolomite. The so-called 'Flexible Limestone', a laminated dolomite bed immediately below the main concretion-bearing horizon, was formerly thought to lie at the base of the Upper Magnesian Limestone but is now known to be well above the base. The Concretionary Limestone passes up and perhaps also laterally into the shallow-water granular and oolitic Hartlepool and Roker Dolomite. Both formations contain species of the bivalves *Schizodus*, *Liebea* and *Permophorus*, and ostracods, gastropods and foraminifera are locally abundant. The Concretionary Limestone has yielded well-preserved holostean fish from

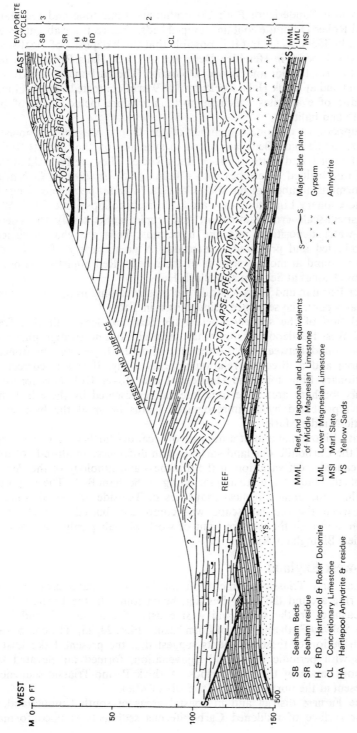

FIG. 26. *Generalized sequence of the Magnesian Limestone in north-east Durham*

SB Seaham Beds
SR Seaham residue
H & RD Hartlepool & Roker Dolomite
CL Concretionary Limestone
HA Hartlepool Anhydrite & residue

MML Reef, and lagoonal and basin equivalents of Middle Magnesian Limestone
LML Lower Magnesian Limestone
MSl Marl Slate
YS Yellow Sands

S——S Major slide plane

⌄ Gypsum

^ Anhydrite

localities near Sunderland. Both Concretionary Limestone and the Hartle-pool and Roker Dolomite contain slumped beds. True reefs are absent because no frame-builders occur, but it is possible that thick deposits of algal-strom-atolites which occur in Heselden Dene and at Black Halls Rocks may form part of the Hartlepool and Roker Dolomite. The carbonates are overlain at Seaham and at Black Halls Rocks by 3 to 9 m of contorted clayey dolomite, the residue of evaporites which in north Yorkshire include beds of both anhydrite and halite.

The uppermost carbonate member of the Upper Magnesian Limestone sequence, the Seaham Beds, are found in synclines on the coast at Seaham and to the north and south of Black Halls Rocks. They are 18 to 32 m thick and are thin-bedded limestones with a fauna of *Schizodus* and *Liebea*, and with enormous numbers of the stick-like alga *Tubulites permianus* King; like the Concretionary Limestone they are extensively collapse-brecciated. Wide-spread small-scale cross-lamination and ripple marks testify to generally shallow-water deposition, and the uppermost beds are characteristically algal-laminated and presumably intertidal. A great variety of calcite con-cretions is found at most exposures of these beds. Red mudstone is present in collapsed zones at Seaham and Black Halls.

Higher Permian and Triassic strata are preserved only in the area south of West Hartlepool, the sequence being shown in Table 3.

Correlation of the sequence below the Seaham Beds with that found north of West Hartlepool is somewhat uncertain, but no stratigraphic break has been found between the Lower and Middle Magnesian limestones in either area and a direct correlation appears likely. If this is correct, the Concretionary Limestone and Hartlepool and Roker Dolomite are absent from this southern area. The time interval represented by these beds may here be represented by an unrecognized hiatus in or at the base of the evaporitic Middle Marls.

Beneath Billingham and at a number of localities farther west, the irregul-arity of the Carboniferous land surface, with differences in altitude of up to 240 m, caused great variation in the thickness and lithology of the Permian sediments that buried it, up to and including the Seaham Beds. The evaporites, upon which the great chemical industries of Teesside are based, were de-posited when the old landscape was completely buried, and show less variation, although the western limit of workable salt partly coincides with the buried Billingham Ridge.

North-west England

Permian and Triassic sediments in north-west England crop out in a broken ring around the Lake District, being found in the Furness district of Lancashire, on the south and west coasts of Cumberland, and in the Carlisle, Solway Firth and Vale of Eden basins (Figs. 24, 25). The composition and distribution of these sediments suggest that the present Lake District, perhaps with periodic uplift and rejuvenation, formed an elevated tract throughout much of the two periods. A thick Permo-Triassic sequence is also present at the northern end of the Isle of Man.

In the **Furness** district and in coastal areas of south Cumberland, the irregular surface of reddened Carboniferous sediments is unconformably

(L 143)

A. Cryptozoon-type algal colonies among stromatolitic reef-flat sediments in the Middle Magnesian Limestone: Hesleden Dene, Durham

Plate VIII

B. Spherulitic calcite in laminated Concretionary Limestone: Cleadon Quarry, South Shields

(L 21/39)

A. Wastwater Screes

Plate IX

B. Wasdale, a major U-shaped valley, and Great Gable

mantled by up to 27 m of red and purple continental breccias ('brockrams') which thicken inland against the flanks of rising ground. They are probably largely of Lower Permian age. Seawards, the breccias are overlain by up to 20 m of grey plant-rich marine or brackish water grey shales and mudstones which contain thin beds of dolomitic limestone. These in turn are succeeded by up to 20 m of buff dolomitic limestone which thins south-westwards to be succeeded by as much as 12 m of anhydrite. This anhydrite contains a network of dolomite, and this is regarded as evidence that it was formed at or slightly above sea level. There can be no doubt that these beds represent a marginal facies of the *Bakevellia* Sea of Zechstein age. The marine strata are overlain by further continental breccias, which flank the generally rising ground and interdigitate successively with up to 200 m of red coastal plain-type siltstones and mudstones—the St. Bees Shales—which are in turn overlain by 760 m of red water-lain St. Bees (Bunter) Sandstone. The Sandstone is succeeded by at least 600 m of grey and red Kirklinton-type sandstones, possibly basal Keuper in age, which in turn are followed by more than 380 m of purple, red and green Keuper Marl. The latter contains abundant sulphates and, under Walney Island, several beds of halite which at one time formed the basis of a small salt industry.

Permian and Triassic sediments are well exposed in coastal sections in **west Cumberland**, where additional information has been derived from many boreholes sunk in connection with coal and anhydrite workings. At Maryport up to 5 m of dolomitic limestone with *Bakevellia* have been proved by upwards borings from undersea coal workings to overlie a few metres of basal sandstones, breccias and mudstones. The most complete sequence in the area has been proved by many borings made near St. Bees Head, the classification of which is as follows:

	Thickness in metres
St. Bees Sandstone	335
St. Bees Shales	91
St. Bees Evaporites, including three carbonate/evaporite cycles	46
Basal Breccia	1·5

The St. Bees Evaporites thin out eastwards against the rising flanks of the Lake District massif, and the St. Bees Shales and lower beds of the St. Bees Sandstone interfinger with and pass eastwards into thick continental breccias banked against the old land surface. The anhydrite of the second of the three evaporite cycles reaches a thickness of 18 m in the west. Anhydrite from the upper two cycles is worked at Sandwith Mine to provide material for the manufacture of sulphuric acid at nearby chemical plants.

As in south Cumberland and Furness, the Permo-Triassic sediments in west Cumberland rest on an uneven reddened erosional surface of Carboniferous sediments and display progressive onlap against rising ground to the east and south. Conversely, the marine Upper Permian strata thicken rapidly westwards into the east Irish Sea sedimentary basins, where thicker and more complete sequences are to be expected.

The sequence at the northern end of the **Isle of Man** is generally similar to those of Cumberland except that it contains no early Upper Permian carbonates or evaporites. Thick Keuper Marl at the top of the sequence contains a number of evaporite beds, including halite.

The **Carlisle Basin** is a broad downwarp centred around the eastern end of the Solway Firth. It is known to contain thick deposits of supposedly Upper Permian and Triassic sediments and, in the south-east, some Lower Permian aeolian sands. Most probably it also contains widespread Upper Permian marine strata, for it seems likely that the evaporites and carbonates of the Vale of Eden trough entered via the Carlisle Basin rather than from the east.

The earliest exposed post-Carboniferous sediment of the Carlisle Basin is breccia, generally less than 1·5 m thick, which is seen at several places around the basin margins. Its relationship to the overlying onlapping St. Bees Shales and Sandstone is similar to that between demonstrably Lower Permian piedmont breccias and Bunter Sandstone in Nottinghamshire and Lincolnshire, and a Lower Permian age must be considered possible at least in part. The breccia is overlain near Eaglesfield and Canonbie in south-west Scotland by apparently aeolian sandstone generally similar to the mainly Lower Permian Penrith Sandstone which crops out in the south-east of the Carlisle Basin and in a narrow subsidiary trough in the valley of the Roe Beck.

The St. Bees Shales of the Carlisle Basin are lithologically similar to their namesake in west and south Cumberland, and pass up by alternation into the thick red St. Bees Sandstone. Gypsum-anhydrite beds are present in the St. Bees Shales at Cocklakes, south-east of Carlisle, and at Riddings, and presumably represent an attenuated marginal marine deposit similar to those of the Vale of Eden. The contact between the St. Bees Sandstone and overlying dull red 90-m Kirklinton Sandstone shows no angular disconformity. The Kirklinton Sandstone in places contains wind-polished grains; it passes up without obvious break into the Stanwix Shales, a thick sequence of red siltstones and mudstones with a few thin beds of dolomitic limestone generally considered to be equivalent to the Keuper Marl elsewhere. Synsedimentary subsidence clearly continued in the Carlisle Basin into Jurassic time, for the Trias is conformably overlain by a large outlier of Liassic strata.

In contrast with that of the Carlisle Basin, the Permian and Triassic sequences of the **Vale of Eden** trough are fairly well known. Due to marked lateral variation, there is no single place where the full sequence is present. A compiled section, representing all beds known to be present, is given in Table 4.

Correlation of the Permian and Triassic strata in the various basins of north-west England is complicated by rapid thickness variations and facies changes, and by diachronous boundaries between formations. The continental breccias, for instance, continued to be deposited in valleys and on the flanks of resistant rocky hills from perhaps late Carboniferous into Triassic time; and movement of sand by wind continued in some places throughout the Lower Permian and may locally have persisted into the Upper Permian, with a revival in the Triassic. Contacts between the sands and breccias

TABLE 4

Permo-Triassic sequence of the Vale of Eden

Formation		Approximate Thickness in metres
St. Bees Sandstone	Bright brick-red water-lain sandstone with many beds of dull red mudstone towards diachronous gradational base. Ripple marks and desiccation cracks present throughout.	300+
Eden Shales	Dull red mudstones and siltstones with subordinate sandstones mainly towards top. Ripple-marks and desiccation cracks present at most levels. Some beds are rich in wind-polished grains and other (mainly grey mudstone) beds are rich in plant remains. Four main gypsum/anhydrite members are widely present in the lower part of the formation; the highest of these is underlain in southern parts of the basin by a thin dolomite yielding a sparse bivalve fauna. The earliest gypsum/anhydrite member is confined to central and southern parts of the basin, and passes laterally into plant-bearing grey, yellow and brown mudstones, siltstones and sandstones (formerly known as the Hilton Plant Beds).	75–180
Penrith Sandstone	Dull brick-red medium- to coarse-grained generally poorly cemented aeolian sandstone, deposited in crescentic dunes by an easterly (relative to modern azimuths) wind. Partly secondarily silicified, especially in the upper part of the formation between Armathwaite and Cliburn. Marginal breccia-wedges and associated water-lain sandstones and mudstones at many levels. Rare vertebrate footprints.	0–460
Brockram	Grey to red ill-sorted continental breccias, composed mainly of fragments of Carboniferous Limestone in a matrix rich in wind-polished sand grains. Partly dolomitized. Represents coalescing desert fans, progressively burying existing steep relief.	

are sharply diachronous, whereas that between the younger St. Bees Shales and St. Bees Sandstone is, over distances of a few kilometres, generally only slightly diachronous. It follows that the various formations to which common names have been given are not necessarily everywhere of the same age.

The marine Upper Permian beds represent incomplete marginal sequences, and full sequences are present only in the deeper parts of the various basins. The flora of the grey beds in the Eden Shales and of the basal grey mudstones and siltstones of the Furness district and St. Bees Head area is similar to that of the basal Zechstein Marl Slate of north-east England, which is almost certainly of partly equivalent age. The limestones and dolomites of the Furness district, as well as those of the earliest cycle in west Cumberland, carry a limited bivalve fauna which is nevertheless characteristic of marginal carbonates of the Lower Magnesian Limestone east of the Pennines; they are overlain by diagenetic sulphates generally similar to the Hartlepool Anhydrite. The principal carbonate of the Vale of Eden, by contrast, bears a sparse fauna more like that of the Upper Magnesian Limestone of south Durham.

8. Jurassic System

The Jurassic Rocks that succeed the Trias near Middlesbrough fall into another regional district—that of East Yorkshire and Lincolnshire. The only remnant left of this great system of rocks in 'Northern England' is to be found near Carlisle (see Fig. 24).

Dark shales with bands of bluish argillaceous limestone form an outlier on the relatively high ground about Great Orton to the west of Carlisle, and have been penetrated in several old wells. Exposures at the surface are poor but are sufficient to prove the rocks to belong to the Lower Lias. There is no direct evidence as to the existence of Rhaetic Beds between the Lias and the Stanwix Shales, but in view of the fact that this formation, though thin, marks a widespread submergence at the end of the Triassic times it may exist beneath the Glacial drift round the Great Orton outlier.

9. Igneous Rocks of Post-Triassic Age

The only definitive example of an igneous rock affecting post-Carboniferous strata in the region is the Armathwaite Dyke. In Cumberland the dyke cuts Permian and Triassic sandstones between Renwick and Dalston, but it has not been located in the Lias west of Carlisle, though this is not surprising as it frequently fails to reach the surface even in drift-free areas. That it links up with a Scottish dyke across the Solway Firth seems certain, for its trace shows clearly on the aeromagnetic map of the estuary, and its continuity with the Cleveland Dyke of Yorkshire is adequately established by the trend of numerous intervening outcrops, though many of these appear to be off-stepped *en échelon*.

The rock is a tholeiite with a devitrified glassy base. Similar tholeiites of the same general trend, namely west-north-west, occur in Northumberland and Durham, the most northern being the Acklington Dyke. Together with the Armathwaite Dyke they are undoubtedly part of a swarm of basic dykes associated with the Tertiary volcanic complex of Mull.

Near Castletown, in the Isle of Man, a group of olivine-dolerite dykes which trend north-west are also believed to be of Tertiary age.

10. Quaternary

The sequence of events in northern England during the Quaternary era, despite the evidence contained in a widespread mantle of drift deposits, is still the subject of lively controversy. Nevertheless, some of the major outlines of the history of the period began to emerge from research during the 1860's and the general picture is gradually becoming clearer as information accumulates and new techniques of study are evolved.

The history of northern England during the preceding Tertiary era is virtually unknown, although several phases of submarine and subaerial peneplanation during the late Miocene and the Pliocene have been postulated by some authors to account for concordant summit levels and topographic benches. It seems certain, however, that the main features of the landscape, including a well-developed drainage system, were already established by the beginning of the Quaternary, and that these features profoundly influenced the subsequent glacial history of the region. Within the last million years England was completely or almost completely covered by great ice sheets on several occasions, the last time being near the end of the so-called Weichselian or Devensian Stage (see Fig. 27) only some 10 to 18 thousand years ago. On the whole, it seems that the broader features of the topography survived the formation and passage of the ice with little change; but the landscape of the region underwent great changes of detail, and almost certainly, overall reduction of relief.

Modification of the mountain areas was mainly by erosion, caused either by ice itself and the accompanying harsh climatic conditions, or by meltwaters. Direct ice erosion of vast areas of the Pennines and Lake District removed virtually all pre-existing soil and unconsolidated deposits, and scraped away great quantities of the underlying rock, leaving it in many places smoothed and polished. It is doubtful whether ordinary valleys buried deeply beneath the ice underwent much erosion by ice at the maxima of the glaciations, but deep cirques or corries formed or enlarged during waning of the last ice sheet are common in and around the Lake District, as are the U-shaped valleys (Plate IXв) so characteristic of valley glaciation. Erosion by meltwater created innumerable drainage channels, some large and spectacular, others sharp and steep. It was thought at one time that most meltwater channels were cut by water escaping from ice-dammed lakes or along ice margins, but many (if not most) are now thought to have been formed by streams flowing into, under, or out of the ice, especially during periods of rapid melting.

Modification of the scenery of the lowland areas included some erosion by ice and meltwater but was mainly accomplished by the deposition of the great volumes of detritus carried from the higher ground. The existing valleys, many of them graded to glacial sea levels at or below 46 m B.O.D., were plugged with thick deposits of clays, sands and gravels, and intervening areas were mantled by extensive sheets of boulder clay. The surface relief of these deposits is generally low, but superb fields of those asymmetrical

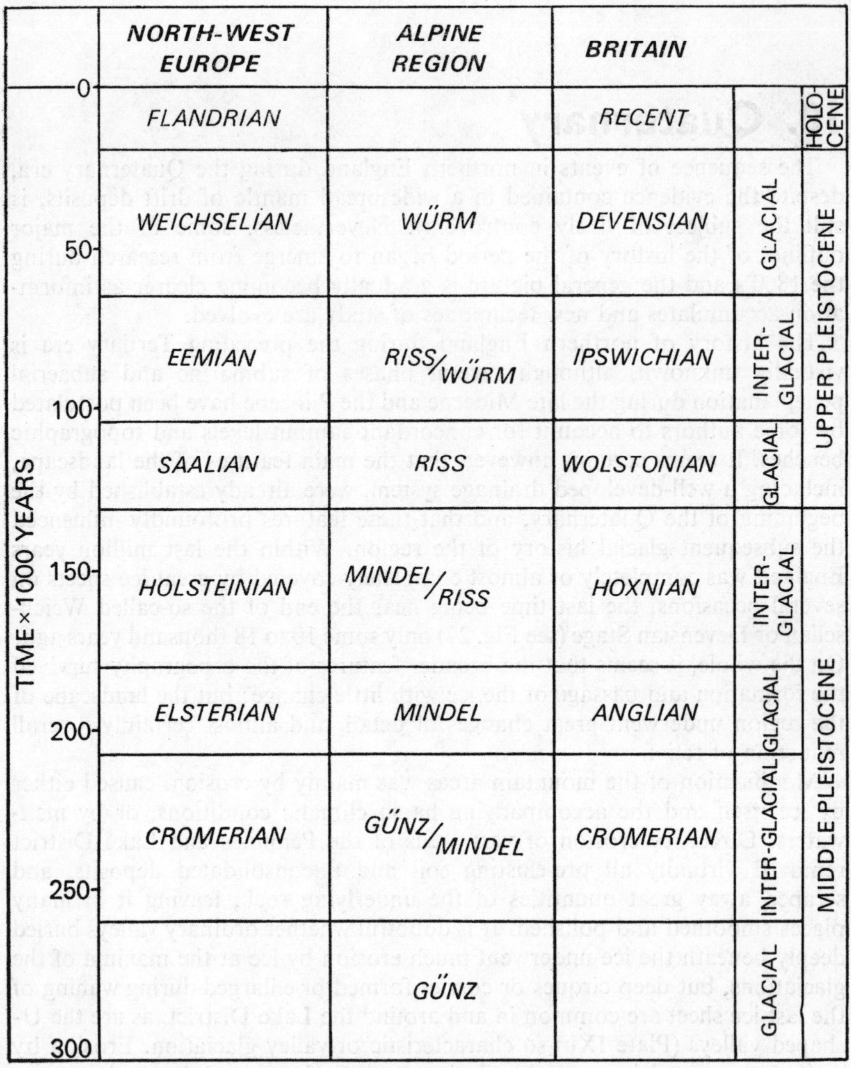

	NORTH-WEST EUROPE	ALPINE REGION	BRITAIN		
0	FLANDRIAN		RECENT		HOLO-CENE
50	WEICHSELIAN	WÜRM	DEVENSIAN	GLACIAL	UPPER PLEISTOCENE
100	EEMIAN	RISS/WÜRM	IPSWICHIAN	INTER-GLACIAL	
	SAALIAN	RISS	WOLSTONIAN	GLACIAL	
150	HOLSTEINIAN	MINDEL/RISS	HOXNIAN	INTER-GLACIAL	
200	ELSTERIAN	MINDEL	ANGLIAN	GLACIAL	MIDDLE PLEISTOCENE
	CROMERIAN	GÜNZ/MINDEL	CROMERIAN	INTER-GLACIAL	
250					
300		GÜNZ		GLACIAL	

(Left axis: TIME × 1000 YEARS)

FIG. 27. *The glacial and interglacial stages of Europe*

rounded mounds of boulder clay known as drumlins are found in several areas where the ice was confined, and distinctive hummocky constructional topography, including eskers and kame and kettle-moraines (Plate XA), occurs in areas where waning ice stagnated.

Surface drainage re-established after the withdrawal of the last ice sheet closely follows the pattern prevailing before the ice arrived, and most of the modern streams lie within or close to an existing drift-filled valley. A few streams, however, were deflected from their original courses; their earlier valleys remain plugged and in some cases are only sketchily known. Marine and fluvial erosion and deposition, closely related to a complicated series of

(*University of Newcastle upon Tyne*)

A. Kettle-moraine at Wooperton, Northumberland.
   Cheviot Hills in distance

Plate X

B. The Trows: glacial meltwater channels cut in Old Red Sandstone andesites near Wooler, Northumberland

(*University of Newcastle upon Tyne*)

changes in the relative level of land and sea, have further modified low-lying areas in the 12 000 or so years since the last ice disappeared.

The Quaternary era is divided into two periods, the Pleistocene and the Holocene (or Recent).

Pleistocene

The earliest known Pleistocene deposits in northern England are fissure-fillings of supposed Middle Pleistocene age in Magnesian Limestone near Blackhall Colliery (County Durham), which have yielded plants, mammalian bones, insects and fresh-water shells. They are believed to be older than the so-called 'Scandinavian Drift', a deposit of grey stony shelly clay which lies in a rock-head hollow at Warren House Gill, some kilometres farther north. This deposit, unique in the region, is somewhat similar to the well-known Basement Clay of Holderness, and may be of Anglian or Wolstonian (Middle or early-Upper Pleistocene) age. The Scandinavian ice sheet appears to have penetrated only a short distance inland from the present coastline, though all other traces of its presence farther inland may well have been removed subsequently.

At Warren House Gill near Horden, County Durham, the Scandinavian Drift is overlain by a thin silty deposit, regarded by C. T. Trechmann as loess and possibly of interglacial origin. Also of interglacial (probably Ipswichian) status is the so-called 'Easington Raised Beach', a deposit of shelly littoral gravel lying on a rock platform at about 27 m above O.D. at Shippersea Bay, 3·2 km north of Warren House Gill. Shells from this gravel have been shown to be older than 38 000 years.

There are no modern descriptions of comparable Ipswichian or earlier deposits west of the Pennines, and the pre-Weichselian history of this area is virtually unknown. It is unlikely, however, that the elevated areas of the whole of the region would have escaped glaciation during the Wolstonian, Anglian and perhaps earlier stages when lowland East Anglia was ice-covered, and it therefore seems that deposits of such glaciations were completely removed or rendered unrecognizable during succeeding interglacials and readvances. A possible relic of pre-Ipswichian glaciations may be a suite of northern England pebbles in the Easington Raised Beach gravel, although there are other explanations of this occurrence. It thus appears that most of Pleistocene time is represented by a few fragmentary deposits on the coast of Durham.

By contrast, the last few thousand years of the Weichselian Stage (from about 18 500 to about 10 300 years ago) are represented by the great spreads of glacial tills and sediments referred to earlier. Study of these sediments has revealed part of the history of an immensely complex glacial episode, in which virtually the whole region was covered by ice at least once and some parts of it perhaps several times. With streams of ice (Fig. 28) emanating simultaneously from centres in Scotland, the Cheviot Hills, the Lake District, and from parts of the Pennines, some lowland areas received ice from more than one centre as the various streams waxed and waned, came together, deflected or rode over each other, or withdrew and readvanced. Typical of such areas are the Carlisle Plain, a receiving ground for ice from north, south and perhaps east, and the Tees lowlands into which ice poured from

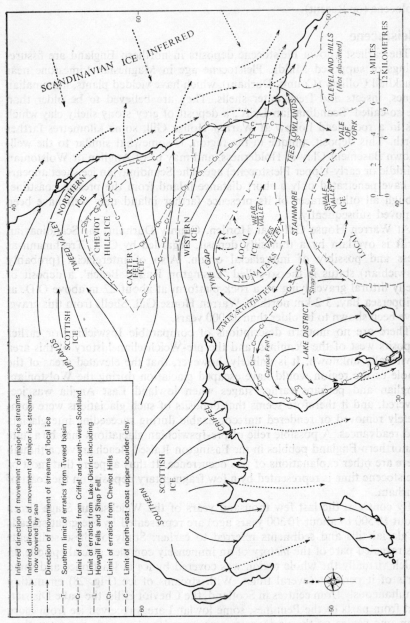

FIG. 28. *The pattern of late-Weichselian (Devensian) ice-movement in northern England*

the west, north-west and north-east. Directions of ice-flow in the coastal areas were greatly influenced by ice offshore, particularly in the north-east where southward deflection of Scottish, Cheviot and Lake District ice is thought to indicate the presence of a mass of Scandinavian ice a few kilometres east of the present coastline. A few of the highest parts of the Pennines are block-strewn and appear to have remained as nunataks even at the height of the late Weichselian glaciation, and the high ground around and east of Cross Fell and on Carter Fell bore local ice which, while fending off the main ice streams, contributed only relatively minor amounts of ice and debris to adjoining lowlands.

West of the Pennines, ice radiating from a thick ice cap on the Lake District mingled with and was partly deflected by great sheets of ice already flowing inexorably southwards from southern Scotland. On the west coast of Cumberland, for instance, drumlinoid till sheets indicate that southward-moving Scottish ice carrying pebbles of Criffel Granodiorite deflected and flowed alongside local Lake District ice carrying erratics from Carrock Fell; similar conflicts occurred in the low ground of the Vale of Eden, where immensely powerful streams of ice moving south-eastwards from Scotland at some stages were deflected eastwards over Stainmore and at others presumably when Lake District ice was particularly thick, flowed eastwards, through the Tyne Gap and over the hilly area to the north. The main deposit formed during this phase was till, representing debris torn from ground overrun by the ice in upstanding areas and subsequently deposited downstream as gradients and flow-rates slackened and the ice became less able to carry its great load. The tills take many forms, ranging from masses of unsorted stones to the most common form, boulder clay. The latter covers large parts of the lower ground, generally resting on rockhead or overlying a few metres of sand and gravel (the latter presumably deposited from meltwater flowing ahead of or beneath the ice), and rising well up the flanks of surrounding higher ground. The surface topography of such sheets commonly reflects, in a subdued way, that of the underlying rock, but drumlin fields are found where the ice moved in constant directions around the margins of the Lake District and in the Tyne Gap. The boulder clay is in most places a hybrid deposit, containing a mixture of far-travelled and local pebbles and taking its colour—generally red or greyish brown—from the softer of the rocks over which it travelled.

Only one till is known in most of north-west England, suggesting that only one major glacial episode occurred in this area in the late Weichselian. It is overlain in places by the products of the waning phases of the ice; these include vast spreads of fluvioglacial sands and gravels deposited both subaerially and subglacially, laminated clays laid down in periglacial lakes, and ridges and mounds of heterogeneous morainic material formed from stagnating ice at the margins of temporary halts and minor readvances during deglaciation. Kame terraces, commonly with well-developed kettle holes and ice-contact slopes, occur in some low-lying areas. This phase of deglaciation is also thought to have been the time when most of the meltwater channels were cut. It is a period when the ice caps dwindled to piedmont glaciers and then to valley glaciers, and when active deepening of some of the Lake District valleys and corries took place.

In a final phase of the glaciation in north-west England, a thin reddish till-like stony clay was deposited in low-lying parts of the Solway Basin and along the coast of north-west Cumberland. The clay swathes pre-existing drumlins and outwash deposits, the tops of which are undisturbed. Its origin and age are uncertain, although it has been interpreted as the till of a minor readvance of Scottish ice. The Bride Hill Moraine near the northern end of the Isle of Man has been taken as marking the limit of this supposed readvance, but has no counterpart in the Solway Basin.

The area east of the Pennines differed from that to the west in that it was mainly a receiving area, only the Cheviot Hills being sufficiently high and extensive to allow the build-up of an ice cap that could form a centre of ice-accumulation. As a result, the north-east was almost completely covered by powerful sheets of ice streaming down the eastern slopes of the hills, and particularly strongly through the Tyne Gap and over Stainmore. The western ice, mainly from south-west Scotland and the Lake District, flowed generally eastwards and southwards, depositing 6 to 9 m of tough brown or grey boulder clay over virtually all parts of Northumberland and Durham and down the Yorkshire coast at least as far as Holderness. Most of this western ice was unconfined, and the boulder clay surface left following its retreat is generally of low relief. Drumlins are well developed, however, where Tweed Basin ice was confined, as in the lower reaches of the Tweed Valley. Ice on the Cheviot Hills and in south-east Scotland appears to have built up more slowly than that in western areas, and was deflected eastwards by western ice already lying to the south.

It is inferred that not far to the east of the present coast lay Scandinavian ice, which deflected southwards successively both the western and northern (Cheviot and south-east Scotland) ice. In the coastal districts it seems that the western ice was initially sufficiently thick and powerful to hold off the northern ice, but as the latter thickened the western ice stagnated and withdrew, allowing northern ice to push a short distance inland and to leave there a reddish brown boulder clay equated with the Hessle Till of Holderness. In its withdrawal, the western ice left behind large spreads of sand and gravel in many low-lying areas along the present coast and in most of the major valleys of Northumberland and Durham; kame and kettle-moraines in north-west and west Durham and parts of Northumberland probably also date from this phase. These deposits (the 'Middle Sands') underlie the boulder clay of the north-eastern ice in coastal districts, and are widely overlain in many of the valleys inland by laminated clays and sands believed to have been laid down in lakes formed when northern ice moved inland and blocked east-ward-flowing drainage courses. Deep channels and associated deltaic deposits in several parts of Durham may have been cut by water overflowing from such lakes. The laminated clays and sands and much of the land up to about 131 m O.D. are covered by as much as 4·6 m of Pelaw Clay, a brown silty sparingly stony clay of unknown origin.

The ice sheets which came together in the Tees lowlands gave rise to at least two reddish brown boulder clays in addition to the almost ubiquitous lower greyish brown clay and to thin intervening deposits of sand, gravel and laminated clay. The reddish brown clays are poorly exposed but at least one of them, perhaps that from the Tees Valley ice, reaches the present coast and may continue south-eastwards to become the Purple Till of Holderness. In all these southern coastal areas, the withdrawal of the western ice eventually allowed east coast ice to sweep well inland, overriding and disrupting laminated clays formed slightly earlier and depositing its characteristic reddish brown boulder clay on top of those derived from the western ice (Plate XIB). Its subsequent stagnation was accompanied in south-east Durham by the deposition of a well-defined kame and kettle-moraine east of Castle Eden, and by the formation of a large lake in which the thick laminated clays and sands of Teesside were laid down. The last major glacial deposit of the lowland areas is the Teesside Clay, a reddish brown stony clay up to 6 m thick which is widely draped over earlier deposits below about 90 m O.D. This clay, like the Pelaw Clay, appears to be a primary deposit, and extends beyond the lacustrine clays and reddish brown tills on to the greyish brown till or even on to rockhead. Its origin is unknown, although deposition from ice floating on the Tees lake appears possible.

After the main ice sheets withdrew, northern England was subjected to a late-glacial phase when periods of intense cold alternated with spells of relatively mild conditions. During the colder spells widespread solifluction modified many of the earlier deposits and led to the formation of extensive 'head' deposits and screes in hilly areas. Modification of existing deposits under permafrost conditions included contortion ('involution') of sands and gravels, well exemplified in coastal cliffs near Whitburn in County Durham and in many inland exposures, and the formation of deep ice-wedges at some localities. Corries in the Lake District and on nearby hills also were actively enlarged during these cold spells, and ice from some of them moved down into adjacent valleys where minor moraines mark maximum extents and temporary pauses during subsequent retreat. Pollen found in swamps and lakes behind some of these moraines show that the last such ice movements took place in Zone III at the end of the Weichselian Stage, about 10 500 years ago.

The climatic history of the late-glacial phase is best illustrated by studies of deposits filling hollows left as the main ice sheets retreated. At several such sites in Cumberland, Northumberland and Durham, cold spells are represented by barren or only scantily fossiliferous sediments, whilst milder spells are recorded in peats and other deposits in which the sequences of pollen and plants eloquently chronicle the gradual re-establishment of a plant cover.

The late-glacial phase is also one in which water formerly locked up in the great ice sheets returned to the oceans, raising sea level rapidly from its glacial level of about 128 m below O.D. Rising sea level, and isostatic recovery of the land which had been weighed down by the mass of ice appear at times to have acted together to produce an apparent stability which lasted long enough to allow the formation of many stream terraces and of supposed marine planations at a number of coastal localities in northern England.

Holocene

Changes in northern England during the 10 500 years of the Holocene period are insignificant compared with those of the preceding 10 000 years. The main changes are undoubtedly related to the final return to present sea level (the 'Flandrian Transgression'), which was associated with continuing shortening and silting up of existing valleys as base levels changed, and with the inundation of former fresh-water swamps as the sea invaded former land areas. The inundation is most strikingly demonstrated in the case of the so-called 'submerged forests', peat deposits which occur at several coastal localities in the Isle of Man, Cumberland, Northumberland and Durham and which at Hartlepool are known to extend to at least 15 m below present sea level. Most of these peats, as well as many basin peats at inland localities and hill peats on the higher ground, were formed in the Boreal and Atlantic stages from about 7500 to 3000 years B.C., when the climate was appreciably milder than today. Bones and antlers of deer are present in many of the peats, and trunks and boles of trees are fairly common in the thicker deposits, Mesolithic flint and bone implements have also been reported. The coastal peats are commonly overlain by recent beach deposits and estuarine alluvium, and are exposed only intermittently; inland peats, likewise, are seldom found at the surface in most lowland areas, where they are generally interbedded with alluvium.

Apart from alluvium, wind-blown sand constitutes the most obvious of the recent deposits around the present coast, but stretches of marine alluvium left behind following a relatively recent sea level fall of 3 to 5 m are present at a few places. Spreads of calcareous tufa are a fairly common recent deposit in areas where limestone or gravels rich in limestone pebbles are present.

(A 6621)

A. Bowscale Tarn, a northward-facing corrie in Skiddaw Slates, dammed by moraine

Plate XI

B. Reverse fault in laminated clay, overlain by undisturbed boulder clay: Sheraton, Co. Durham

(*P. Beaumont*)

Plate XII   Opencast coal pit near Widdrington, Northumberland. Four seams are being worked. Restored ground to left of village

(*Derek Crouch (Contractors) Ltd.*)

11. Economic Geology

Coal has been of prime importance in the industrial growth of the region. In Cumberland all the coals are in the bituminous class, suitable for steam-raising, coking and household use. A few centimetres of cannel, which may pass laterally into carbonaceous shale, is present above some of the seams. The thinness of most of the seams, and the extensive and heavy faulting, have always made coal-mining difficult in Cumberland. Most of the pits, and all the inland ones, have now been closed, either because they were exhausted or because they became uneconomic. Some of the remaining coal on land is being worked by opencast methods, but the main prospect lies offshore. Here the geological difficulties are as great as on land and ventilation problems are increased as the workings move farther from land-based shafts. In spite of this there are still (1970) two active collieries—Solway and Haig pits, the workings in the latter having now reached some 6·5 km from the coast.

In Northumberland and Durham the 250 or so metres above and including the Brockwell Seam contain all the best coals, though these vary greatly in thickness and quality. They include the Busty, Harvey, Bensham and High Main seams, the latter at one time being so well known as a good house coal that its name was locally given to other inferior seams for commercial purposes, leading to some confusion in stratigraphical nomenclature.

The variation in the character of the coal in any one seam across the area is greater than the variation between different seams at any one locality, and it is therefore possible to divide the coalfield into broad areas in accordance with the coal type. These are shown on Fig. 22. Generally speaking, coal of the highest rank occurs in west Durham, where it produces an exceptionally good coke, low in ash, sulphur and phosphorus. It is used for both foundry and general metallurgical work and its occurrence was one of the reasons for the development of the steel industry at Consett. To the north, east and south the volatile content increases, and there is a progressive decrease in carbon content and calorific value. Modern methods of blending and treatment of solid fuel render these variations of less importance than they once were.

As in Cumberland, most of the inland pits are now either exhausted or uneconomic, and deep mining of coal is becoming concentrated in a handful of large pits near the coast, exploitation in the rest of the coalfield being largely by opencast methods.

Coals in the Viséan and Namurian, of which the Shilbottle and Little Limestone coals are the most important (see Plates V and VII) are still mined on a small scale in Northumberland.

Rock salt occurs in the Keuper Marl on Walney Island and at the northern end of the Isle of Man, and natural brine from the solution of these beds has in the past been produced at both places. Today, production is centred around Greatham in south-east Durham, where artificially induced brine is pumped from solution cavities in a salt bed up to 45 m thick in the Permian Middle Evaporite Group. This brine is an essential raw material in the important heavy chemical industry of Teesside, where it is the basis of the alkali trade and of the manufacture of chlorine and its compounds. Many other branches of industry depend in turn on these primary products, and reserves are ample for many decades at the present rate of production. A small amount of table and preserving salt is produced by the evaporation of brine, but this trade is declining in Teesside.

Gypsum and anhydrite, the hydrous and anhydrous forms, respectively, of calcium sulphate, occur as bedded deposits in the Permian rocks of west Cumberland, the Vale of Eden, and south-east Durham. Generally speaking, gypsum occurs at shallow depths near the outcrop, and passes into anhydrite in depth. Gypsum is chiefly used to make plaster and plaster-board and as a retarder in Portland cement, and it has several other minor uses. Anhydrite is important in the manufacture of ammonium sulphate fertilizer and of sulphuric acid—the latter process yielding an important by-product in the form of cement.

Gypsum occurs in the Eden Shales in the Vale of Eden, between Cotehill, south-east of Carlisle, and Kirkby Thore, north-west of Appleby, and has been worked by both underground and opencast methods. Some anhydrite is also won here. A mine in the thick anhydrite beds in the St. Bees Evaporites beneath St. Bees Head supplies a nearby sulphuric acid and cement works, and the Billingham Main Anhydrite on Teesside is likewise mined and used in the Billingham chemical plants.

Over 2 million tons of anhydrite and gypsum are produced annually in the region, representing the total national output of anhydrite and about one-fifth of that of gypsum. Sulphuric acid based on anhydrite suffers some price competition with that produced from imported sulphur. Any decline in anhydrite production will therefore probably be due to this economic cause, rather than to a shortage of reserves, which are very great. Demand for gypsum products, on the other hand, is expected to go on increasing, and it is unfortunate that its mode of occurrence near to the outcrop means that its reserves are probably limited to a small fraction of those of anhydrite.

Limestone is among the most important of the bulk minerals within the region, and total production rose steadily from 2·3 million tons in 1948 to more than 6 million tons in 1969. It is used in cement and lime production, as roadstone and railway ballast, and in powdered form for agricultural purposes. Limestone of high chemical purity is used as a flux in the iron and steel works of Consett and Teesside. The thick limestone beds of the Carboniferous are the main source. Most of these are in the Lower Carboniferous, but the Namurian Great Limestone is also important in Weardale where it attains a thickness of some 22 m and provides the raw material for a large cement works, among other uses.

The Permian Magnesian Limestone of east Durham ranges in composition from fairly pure calcium carbonate to high-grade dolomite, and some of its uses overlap with those of limestones from the Carboniferous. Other aspects and uses which depend on the magnesium content are described in the section on dolomite, below.

There is an increasing demand for hard limestone for aggregate in concrete, and whilst many of the Carboniferous limestones are suitable for this purpose it is necessary to be highly selective in the case of the Magnesian Limestone. Hard limestones with high crushing strengths are to be found in the Lower and Upper Magnesian Limestones; the Middle Magnesian Limestone has a tendency to be porous and to lack strength, except where it is part of the reef facies. There is some demand for ground limestone as a filler or a mild abrasive. For instance, a large quarry at Ford near Sunderland has supplied Magnesian Limestone which, when powdered, was used in the manufacture of tooth-paste and cosmetics.

Dolomite is of great importance in the iron and steel industry, where it is used for making refractory bricks for furnace and converter linings in basic steel-making processes. It is also used in the pharmaceutical, glass-making, tanning and textile industries. Production in Durham rose from 0·9 million tons in 1949 to 2·1 million tons in 1965, and accounted for more than half the total output for England and Wales. Reserves of all forms of limestones are very large. Parts of the Magnesian Limestone in Durham are true dolomites, and contain more than 40 per cent of magnesium carbonate. They are worked in several quarries between South Hylton and Aycliffe, the main concentration being between Coxhoe and Mainsforth.

In the field of **refractories,** some of the Carboniferous sandstones, particularly in west Durham, have a high enough silica content to be worked as ganisters for use in the manufacture of firebricks and furnace linings. Some beds, lacking cohesion between the grains, are worked as moulding sands near Stanhope. Other less refractory moulding sands are obtained from glacial deposits, and, at one place, from the bed of the Tees. Reserves are thought to be adequate, and the Permian Yellow Sands might be considered as another source of moulding sand; they are already used for making silica bricks.

Fireclay, which is generally associated with coal seams, is got from both surface and underground workings in the Lower and Middle Coal Measures of Cumberland, Durham and Northumberland, and in the Millstone Grit Series of the Hexham and Corbridge area. It is used in the manufacture of refractory goods and sanitary ware. Output in the region was nearly 150 thousand tons in 1965, and reserves are adequate for many years.

Brick clays are of very wide occurrence in the region. They include Coal Measures shale, such as is worked in Lumbley and Pelaw in Durham, at Ashington and Throckley in Northumberland and at Whitehaven and Harrington in Cumberland. The Pleistocene laminated clay which covers large areas of east Durham and Northumberland is also widely used for brickmaking, and Newcastle and Sunderland were largely built of bricks made from it. Boulder clay is ubiquitous on the lower ground on both sides of the Pennines. It is the raw material for most of the bricks made in Cumberland, and there are several pits in it in the Carlisle area. At Felling, near Gateshead, Coal Measures shale has been worked and processed together with the overlying boulder clay, and at Greenscoe in Furness mudstones in the Skiddaw Group are worked for brickmaking.

The use of natural **building stone** has been superseded to some extent by other materials, but there remains a steady demand for it, not only for constructional work, but as kerbs, flags, garden stone, and for ornamental purposes. Sources of building stone are so numerous and reserves so great that no more than a few of them can be mentioned here. Any reasonably durable stone, including boulders from the drift, can be used for rough walling, and it is only necessary to consider rocks of some commercial importance.

Starting with the Lower Palaeozoic, the Skiddaw Slates have been used locally for building, but of much greater importance are the cleaved tuffs of the Borrowdale Volcanic Group, which especially when sawn, provide a handsome greenish grey stone for building, facing, paving and ornamental uses. These rocks also yield heavy roofing slates.

In the Carboniferous, durable sandstones are found at many horizons, and most of the villages and farms, and the older parts of towns on the outcrop are built of sandstone derived from small local quarries. The characteristic purplish red colour of the Whitehaven Sandstone is seen in many of the buildings in towns along the west coast, and in Northumberland and Durham numerous fine churches and public buildings, including Durham Cathedral, bear witness to the durable and freeworking qualities of some of the thick sandstone 'posts' of the Coal Measures.

Among the Permo-Triassic rocks, in the west of the region the harder and more quartzitic parts of the Penrith Sandstone are used to some extent for building, but the reddish brown St. Bees Sandstone is of prime importance in west Cumberland and on the Solway Plain. In Durham, the Permian Magnesian Limestone once had a limited use for building, but its main use is now in the metallurgical and chemical fields.

Among the igneous rocks the tough Shap Granite stands out as an extremely durable stone which is used extensively in building bridges, impounding dams, and docks—the King George V Graving Dock at Southampton is an example of the latter. The polished rock, with its large pink feldspar phenocrysts, is a familiar facing stone on public buildings throughout the country.

Sources of rock which can be crushed for **roadstone** or **concrete aggregate** include igneous rocks such as the Threlkeld and Shap granites, the Whin Sill and the Cleveland and Hett dykes. The harder limestones are also used, as are gravels derived by glacial outwash from ice that has passed over hard rocks. Almost any reasonably durable stone, and this includes many sandstones whose strengths are inadequate for use as aggregates, can be pressed into service locally for road ballast.

As in other parts of the country, the demand for **sand and gravel,** mainly for use in concrete, has greatly increased in recent years. Production in the region was over 2 million cu m in 1965 and the demand is expected to double in about 10 years. A large proportion of the output is from the glacial outwash deposits of the last (Weichselian) ice sheet, which mantle most of the lower ground. In Northumberland and Durham the quality is very variable. Clay content can be high, and fragments of coal, and in east Durham, Magnesian Limestone, lower its value as a building material. Quality is better west of the Pennines, and there are large pits in the northern part of the Vale of Eden, and on the Solway Plain near Carlisle. Much of the remainder of the output comes from alluvial terrace gravels of the rivers Tyne, Wear and Tees. Small amounts of sand and gravel are dug on the sea shore in Northumberland and Durham, and gravel is obtained in small quantities from the beds of rivers in Cumberland and Northumberland, and from Lake Windermere. The relatively soft and well-graded Permian Yellow Sands are quarried for building sand near Sherburn Hill, Eppleton and Houghton le Spring in Durham, and there is much scope for expansion.

The sideritic clay ironstones of the Coal Measures in Durham and Cumberland, together with the adjacent coal seams, formed the basis of an early iron industry. Today, the **hematite** of west Cumberland is the only iron ore worked in the region. It ranks among the richest ores in the country, averaging 48 per cent iron content, and providing the main domestic source of low phosphorus ore. The hematite occurs as replacement deposits in the Carboniferous limestones, in the form of veins, or vein-like bodies along fault lines; as 'flats'—tabular bodies extending laterally in favourable beds of limestones; and as 'sops' which have the form of an inverted cone with irregular walls. Some limestones have lent themselves more readily than others to replacement by hematite.

All the shallower ore has now been mined, and out of many mines once producing in Cumberland and Furness only three, all in the Egremont area, are still active. Production has declined from the maximum of 1·5 million tons in the 1880's to 0·7 million tons in 1938, thence to about 0·27 million tons at the present time, representing about 65 per cent of the national output of hematite. In places sufficient manganite and other manganese minerals have been present to make ore bodies workable for their manganese content. Reserves of hematite in Cumberland are limited, and the chances of discovering new large ore bodies in situations where they could be economically worked are not thought to be high.

Non-ferrous ores occur widely in the region, particularly in the Carboniferous rocks of the north Pennines, and in the Ordovician Skiddaw Slates of the Lake district. Emplacement is in the form of veins along fissures and as 'flats', in which limestone has been replaced laterally by ore-bearing solutions emanating from adjacent fissures. Minerals of economic value include galena and zinc blende—sulphides of lead and zinc respectively—and the associated spar minerals, fluorspar, barytes and witherite.

In recent years production of the ores of lead and zinc has declined almost to nothing, except as a by-product of the mining of the spar minerals. **Fluorspar** is in demand for the manufacture of hydrofluoric acid and fluorine compounds, and as a flux in steel-making. County Durham produces about one-quarter of the total for England and Wales. **Barytes** and **witherite** are used as high density fillers in the paint and paper industries and as a source of barium chemicals. There are grounds for believing that economic reserves of both metal ores and associated minerals may exist, particularly in the north Pennines, but much exploratory work would be needed to establish this.

In the Kielder–Cheviot area there are many disconnected patches of upland blanket **peat** which coalesce on the highest ground to form considerable areas up to 2·5 m thick. The widest spreads of blanket peat exist in the northern Pennines where, but for the main valleys which interrupt it, there would be a more or less continuous sheet of peat for some 120 km between Lambley and the moors above Skipton, far to the south of the region. Much of this peat is up to 2·5 m thick but the present tendency of English upland peat is to waste away, and to be represented over wide areas by 'peat hags', between which the deposit is thin or absent. Numerous runnels dissect the peat still further. There are no very large spreads in the Lake District, the most extensive being in the Black Combe–Whitfell and Shap Fells areas; elsewhere it appears that hill slopes are too steep to support or retain peat. There are no commercial workings of upland peat though it is cut for burning at a few remote farms. Its main, though indirect, economic value lies in its ability to absorb water in heavy storms releasing it gradually to the river catchments, thus helping to a small extent to prevent the sudden dangerous flash floods which tend to occur in the narrow upland stream valleys.

The largest areas of lowland peat are found in the Solway Plain. Both Solway Moss, near Longtown, and Wedholme Flow, south of Kirkbride, have been extensively worked for moss litter, and there are other smaller workings. There are considerable reserves in the area.

The siliceous fossil remains of microscopic organisms occur as an early post-glacial sediment, **diatomite**, on the bed of a few lakes and tarns in the Lake District, of which Kentmere is the best known. Here the diatomite is worked chiefly for insulation purposes.

In the field of **water supply** certain beds of rock are important as aquifers. In the Carboniferous these include most of the thick gritstones and sandstones, and some of the limestones, but yields are often unpredictable since the permeability is due to the incidence of joints and fissures in the brittle close-grained rock. In the Permian, the Penrith Sandstone and the Yellow Sands of Durham are excellent and reliable aquifers because of their high granular porosity, and the Middle Magnesian Limestone also yields large amounts of water in south-east Durham, though permanent hardness is high.

There is much scope within the region for the practice of **engineering geology**. Incipient landslips, mining subsidence, the incidence of clays of low load-bearing capacity along the coastal plain in the east, are all examples of problems that are encountered in the building of roads and heavy structures. The design of impounding dams requires a detailed knowledge of the local geology and of available constructional materials.

12. Geological Survey Maps and Memoirs, and Select Bibliography of other works

Sheet and District Memoirs

These are all Memoirs of the Geological Survey of Great Britain, and are published by Her Majesty's Stationery Office, London. Some of the older works are out of print. The numbers in brackets refer to the One-inch New Series maps of the area described.

(1, 2)   FOWLER, A. 1926. The geology of Berwick on Tweed, Norham and Scremerston. ix+58 pp.

(3, 5)   CARRUTHERS, R. G., BURNETT, G. A. and ANDERSON, W. 1932. The geology of the Cheviot Hills. xii+174 pp.

(4)   —— DINHAM, C. H., BURNETT, G. A. and MADEN, J. 1927. The geology of Belford, Holy Island and the Farne Islands. xi+195 pp.

(6)   —— BURNETT, G. A. and ANDERSON, W. 1930. The geology of the Alnwick district. xiii+138 pp.

(7)   CLOUGH, C. T. 1889. The geology of Plashetts and Kielder. 68 pp.

(8)   MILLER, HUGH. 1887. The geology of the country between Otterburn and Elsdon. viii+147 pp.

(9, 10)   FOWLER, A. 1936. The geology of the country around Rothbury, Amble and Ashington. xii+159 pp.

(11, 16, 17)   DIXON, E. E. L., MADEN, J., TROTTER, F. M., HOLLINGWORTH, S. E. and TONKS, L. H. 1926. The geology of the Carlisle, Longtown and Silloth district. xiii+113 pp.

(12)   DAY, J. B. W. 1970. Geology of the country around Bewcastle. xi+357 pp.

(15)   LAND, D. H. In preparation. Geology of the Tynemouth district.

(18)   TROTTER, F. M. and HOLLINGWORTH, S. E. 1932. The geology of the Brampton district. xviii+223 pp.

(22)   EASTWOOD, T. 1930. The geology of the Maryport district. xiii+137 pp.

(23)   —— HOLLINGWORTH, S. E., ROSE, W. C. C. and TROTTER, F. M. 1968. Geology of the country around Cockermouth and Caldbeck. x+298 pp.

(24)   ARTHURTON, R. S. and WADGE, A. J. In preparation. Geology of the Penrith district.

(27)   SMITH, D. B. and FRANCIS, E. A. 1967. Geology of the country between Durham and West Hartlepool. xiii+354 pp.

(28)   EASTWOOD, T., Dixon E. E. L., HOLLINGWORTH, S. E. and SMITH, B. 1937. The geology of the Whitehaven and Workington district. xvii+304 pp.

(30)   DAKYNS, J. R., TIDDEMAN, R. H. and GOODCHILD, J. G. 1897. The geology of the country between Appleby, Ullswater and Haweswater. vi+110 pp.

(32) MILLS, D. A. C. and HULL, J. H. In preparation. Geology of the country around Barnard Castle.

(37) TROTTER, F. M., HOLLINGWORTH, S. E., EASTWOOD, T. and ROSE, W. C. C. 1937. Gosforth District. xii+136 pp.

(39) AVELINE, W. T. and HUGHES, T. McK. 1888. The geology of the country around Kendal, Sedbergh, Bowness and Tebay; Second edition revised and enlarged, by A. Strahan. vii+944 pp.

(40) DAKYNS, J. R., TIDDEMAN, R. H., RUSSELL, R., CLOUGH, C. T. and STRAHAN, A. 1891. The geology of the country around Mallerstang; with parts of Wensleydale, Swaledale and Arkendale. x+213 pp.

(49) AVELINE, W. T., HUGHES, T. McK. and TIDDEMAN, R. H. 1872. The geology of the neighbourhood of Kirkby Lonsdale and Kendal. 44 pp.

(58) —— 1873. Geology of the southern part of the Furness district. 13 pp.

(I.O.M.) LAMPLUGH, G. W. 1903. The geology of the Isle of Man. xiv+620 pp.

WARD, J. C. 1876. The geology of the northern part of the English Lake District. xii+132 pp.

DUNHAM, K. C. 1948. Geology of the Northern Pennine orefield. vi+357 pp.

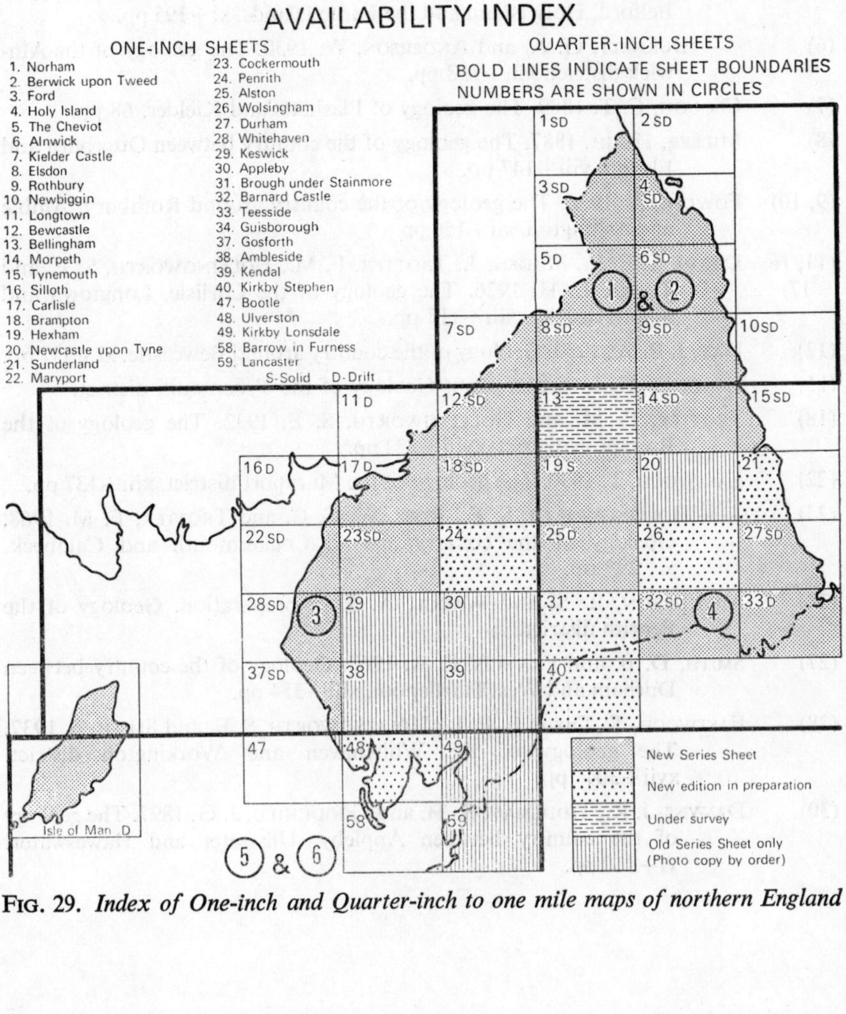
AVAILABILITY INDEX
ONE-INCH SHEETS
1. Norham
2. Berwick upon Tweed
3. Ford
4. Holy Island
5. The Cheviot
6. Alnwick
7. Kielder Castle
8. Elsdon
9. Rothbury
10. Newbiggin
11. Longtown
12. Bewcastle
13. Bellingham
14. Morpeth
15. Tynemouth
16. Silloth
17. Carlisle
18. Brampton
19. Hexham
20. Newcastle upon Tyne
21. Sunderland
22. Maryport
23. Cockermouth
24. Penrith
25. Alston
26. Wolsingham
27. Durham
28. Whitehaven
29. Keswick
30. Appleby
31. Brough under Stainmore
32. Barnard Castle
33. Teesside
34. Guisborough
37. Gosforth
38. Ambleside
39. Kendal
40. Kirkby Stephen
47. Bootle
48. Ulverston
49. Kirkby Lonsdale
58. Barrow in Furness
59. Lancaster
S-Solid D-Drift
QUARTER-INCH SHEETS
BOLD LINES INDICATE SHEET BOUNDARIES
NUMBERS ARE SHOWN IN CIRCLES
1 SD
2 SD
3 SD
4 SD
5 D
6 SD
7 SD
8 SD
9 SD
10 SD
11 D
12 SD
13
14 SD
15 SD
16 D
17 D
18 SD
19 S
20
21
22 SD
23 SD
24
25 D
26
27 SD
28 SD
29
30
31
32 SD
33 D
37 SD
38
39
40
47
48
49
58
59
Isle of Man D
New Series Sheet
New edition in preparation
Under survey
Old Series Sheet only
(Photo copy by order)

FIG. 29. *Index of One-inch and Quarter-inch to one mile maps of northern England*

General

ANSON, W. W. and SHARP, J. J. 1960. Surface and rock-head relief features in the northern part of the Northumberland coalfield. *University of Durham (Kings College) Dept. of Geography*. Research Series, No. 2.

EVANS, J. W. and STUBBLEFIELD, C. J. 1929. *Handbook of the geology of Great Britain*. xii+556 pp. Murby, London.

HOLLINGWORTH, S. E. 1954. The Geology of the Lake District. *Proc. Geol. Ass.*, **65**, 385–402.

JOHNSON, G. A. L. and DUNHAM, K. C. 1963. The Geology of Moor House. *Monogr. Nat. Conserv.*, No. 2, 182 pp. H.M.S.O., London.

JONES, J. M. 1967. Geology of the coast section from Tynemouth to Seaton Sluice. *Trans. nat. Hist. Soc. Northumb.*, **16** (New Series), 153–92.

KENT, P. E. 1966. The structure of the concealed Carboniferous rocks of north-eastern England. *Proc. Yorks. geol. Soc.*, **35**, 323–52.

MARR, J. E. 1916. *Geology of the Lake District*. xii+220 pp. Cambridge.

MITCHELL, G. H. 1956. The geological history of the Lake District. *Proc. Yorks. geol. Soc.*, **30**, 407–63.

RIDD, M. F., WALKER, D. B. and JONES, J. M. 1970. A deep borehole at Harton on the margin of the Northumbrian Trough. *Proc. Yorks. geol. Soc.*, **38**, 75–103.

SHIELLS, K. A. G. 1964. The geological structure of north-east Northumberland. *Trans. R. Soc. Edinb.*, **65**, 447–81.

SHOTTON, F. W. and TROTTER, F. M. 1936. Cross Fell and Stainmore (Report of field meeting). *Proc. Geol. Ass.*, **47**, 376–87.

SMITH, B. and OTHERS. 1925. Sketch of the geology of the Whitehaven District. *Proc. Geol. Ass.*, **26**, 37–75.

TROTTER, F. M. and HOLLINGWORTH, S. E. 1928. The Alston Block. *Geol. Mag.*, **65**, 433–48.

TURNER, J. S. 1935. Structural geology of Stainmore, Westmorland and notes on the Late Palaeozoic (late-Variscan) tectonics of the north of England. *Proc. Geol. Ass.*, **46**, 121–51.

—— 1949. The deeper structure of central and northern England. *Proc. Yorks. geol. Soc.*, **27**, 280–97.

VERSEY, H. C. 1960. *A guide to the geology of the Appleby District*. 40 pp. Whitehead, Appleby.

WESTOLL, T. S. and OTHERS. 1955. A guide to the geology of the district around Alnwick. *Proc. Yorks. geol. Soc.*, **30**, 61–100.

Ordovician and Silurian

BROWN, P. E., MILLER, J. A. and SOPER, N. J. 1964. Age of the principal intrusions of the Lake District. *Proc. Yorks. geol. Soc.*, **34**, 331–42.

CLARK, L. 1964. The Borrowdale Volcanic series between Buttermere and Wasdale, Cumberland. *Proc. Yorks. geol. Soc.*, **34**, 343–56.

DEAN, W. T. 1959. The Stratigraphy of the Caradoc Series in the Cross Fell Inlier. *Proc. Yorks. geol. Soc.*, **32**, 185–227.

DOWNIE, C. and FORD, T. D. 1966. Microfossils from the Manx Slate Series. *Proc. Yorks. geol. Soc.*, **35**, 307–22.

ELLES, G. L. 1898. Graptolite Faunas of the Skiddaw Slates. *Q. J. geol. Soc. Lond.*, **54**, 463–539.

FIRMAN, R. J. 1957. Borrowdale Volcanic Series between Wastwater and Duddon Valley. *Proc. Yorks. geol. Soc.*, **31**, 39–64.

FURNESS, R. R. 1965. The petrography and provenance of the Coniston Grits east of the Lune Valley, Westmorland. *Geol. Mag.*, **102**, 252–60.

—— LEWELLYN, P. G., NORMAN, T. N. and RICKARDS, R. B. 1967. A review of Wenlock and Ludlow stratigraphy and sedimentation in N.W. England. *Geol. Mag.*, **104**, 132–47.

GREEN, J. F. N. 1919. Vulcanicity of the Lake District. *Proc. Geol. Ass.*, **30**, 153–82.

HARTLEY, J. J. 1925. Borrowdale Volcanic Series between Windermere and Coniston. *Proc. Geol. Ass.*, **36**, 203–36.

—— 1932. The volcanic and other igneous rocks of Great and Little Langdale. *Proc. Geol. Ass.*, **43**, 32–69.

—— 1942. The geology of Helvellyn and the southern part of Thirlmere. *Q. J. geol. Soc. Lond.*, **97**, 129–62.

HELM, D. G. 1970. Stratigraphy and structure in the Black Coombe Inlier, English Lake District. *Proc. Yorks. geol. Soc.*, **38**, 105–48.

INGHAM, J. K. 1966. The Ordovician rocks in the Cautley and Dent districts of Westmorland and Yorkshire. *Proc. Yorks. geol. Soc.*, **35**, 455–505.

—— and WRIGHT, A. O. 1970. A revised classification of the Ashgill Series. *Lethia*, **3**, 233–42.

JACKSON, D. E. 1961. Stratigraphy of the Skiddaw Group between Buttermere and Mungrisdale, Cumberland. *Geol. Mag.*, **98**, 515–28.

—— 1962. Graptolite zones in the Skiddaw Group in Cumberland, England. *J. Palaeont.*, **36**, 300–13.

KING, W. B. R. and WILLIAMS, A. 1948. On the Lower Part of the Ashgillian Series in the North of England. *Geol. Mag.*, **85**, 205–12.

MARR, J. E. 1913. The Lower Palaeozoic rocks of the Cautley district. *Q. J. geol. Soc. Lond.*, **69**, 1–17.

—— and NICHOLSON, H. A. 1888. The Stockdale Shales. *Q. J. geol. Soc. Lond.*, **44**, 654–732.

MITCHELL, G. H. 1940. Borrowdale Volcanic Series of Coniston, Lancashire, *Q. J. geol. Soc. Lond.*, **46**, 301–319.

—— 1963. The Borrowdale Volcanic rocks of the Seathwaite Fells. *Lpool. Manchr. geol. J.*, **3**, 289–99.

MOSELEY, F. 1960. The succession and structure of the Borrowdale Volcanic rocks south-east of Ullswater. *Q. J. geol. Soc. Lond.*, **116**, 55–84.

—— 1964. The succession and structure of the Borrowdale volcanic rocks north-west of Ullswater. *Geol. J.*, **4**, 127–42.

NICHOLSON, H. A. and MARR, J. E. 1891. Cross Fell Inlier. *Q. J. geol. Soc. Lond.*, **47**, 500–29.

NUTT, M. J. C. 1966. Field meeting report. *Proc. Yorks. geol. Soc.*, **35**, 429–33.

OLIVER, R. L. 1961. The Borrowdale volcanic and associated rocks of the Scafell area, Lake District. *Q. J. geol. Soc. Lond.*, **117**, 377–417.

RICKARDS, R. B. 1964. The graptolitic mudstone and associated facies in the Silurian strata of the Howgill Fells. *Geol. Mag.*, **101**, 435–51.

—— 1967. The Wenlock and Ludlow succession in the Howgill Fells (north-west Yorkshire and Westmorland). *Q. J. geol. Soc. Lond.*, **123**, 215–51.

SIMPSON, A. 1963. The stratigraphy and tectonics of the Manx Slate Series, Isle of Man. *Q. J. geol. Soc. Lond.*, **119**, 367–400.
—— 1967. The stratigraphy and tectonics of the Skiddaw Slates and the relationship of the overlying Borrowdale Volcanic Series in part of the Lake District. *Geol. J.*, **5**, 391–418.

SHOTTON, F. W. 1935. The Stratigraphy and Tectonics of the Cross Fell Inlier. *Q. J. geol. Soc. Lond.*, **91**, 639–704.

SOPER, N. J. 1970. Three critical localities on the junction of the Borrowdale Volcanic rocks with the Skiddaw Slates in the Lake District. *Proc. Yorks. geol. Soc.*, **37**, 461–93.

Carboniferous

ARMSTRONG, G. and PRICE, R. H. 1954. The Coal Measures of north-east Durham. *Trans. Instn. Min. Engrs.*, **113**, 974–97.

CALVER, M. A. 1968. Distribution of the Westphalian marine faunas in northern England and adjoining areas. *Proc. Yorks. geol. Soc.*, **37**, 1–72.

CARRUTHERS, R. G. 1938. Alston Moor to Botany and Tanhill: an adventure in stratigraphy. *Proc. Yorks. geol. Soc.*, **23**, 236–53.

DUNHAM, K. C. 1950. Lower Carboniferous sedimentation in the northern Pennines (England). Rept. XVIII. *Int. Geol. Congr.*, Pt 4, 46–63.
—— and ROSE, W. C. C. 1941. The geology of the iron-ore field of south Cumberland and Furness. *War-time Pamphlet Geol. Surv. Gt Br.*, No. 16.
—— and JOHNSON, G. A. L. 1962. Sub-surface data on the Namurian strata of Allenheads, south Northumberland. *Proc. Yorks. geol. Soc.*, **33**, 235–54.

FOWLER, A. 1966. The stratigraphy of the North Tyne basin around Kielder and Falstone, Northumberland. *Bull. geol. Surv. Gt Br.*, No. 24, 57–104.

FROST, D. V. 1969. The Lower Limestone Group (Viséan) of the Otterburn district, Northumberland. *Proc. Yorks. geol. Soc.*, **37**, 277–309.

GARWOOD, E. J. 1913. The Lower Carboniferous succession in the north-west of England. *Q. J. geol. Soc. Lond.*, **68**, 449–586.
—— 1916. The faunal succession in the Lower Carboniferous rocks of Westmorland and north Lancashire. *Proc. Geol. Ass.*, **27**, 1–43.

GEORGE, T. N. 1958. Lower Carboniferous palaeogeography of the British Isles. *Proc. Yorks. geol. Soc.*, **31**, 227–318.

HEDLEY, W. P. 1931. The Stratigraphy of the Bernician and Millstone Grit of south Northumberland. *Trans. nat. Hist. Soc. Northumb.* **7**, (New Series), 179–90.
—— and WAITE, S. T. 1929. The sequence of the Upper Limestone Group between Corbridge and Belsay. *Proc. University of Durham Phil. Soc.*, **8**, 136–52.

HOPKINS, W. and BENNISON, G. M. 1957. A palaeontological link between the Midgeholme Outlier, Cumberland and the Northumberland and Durham coalfield. *Geol. Mag.*, **94**, 215–20.

HULL, J. H. 1968. The Namurian stages of north-eastern England. *Proc. Yorks. geol. Soc.*, **36**, 297–308.

JOHNSON, G. A. L. 1959. The Carboniferous stratigraphy of the Roman Wall district in western Northumberland. *Proc. Yorks. geol. Soc.*, **32**, 83–130.
—— 1967. Basement control of Carboniferous sedimentation in northern England. *Proc. Yorks. geol. Soc.*, **36**, 175–94.
—— HODGE, B. L. and FAIRBAIRN, R. A. 1962. The base of the Namurian and of the Millstone Grit in north-eastern England. *Proc. Yorks. geol. Soc.*, **33**, 341–62.

LEWIS, H. P. 1930. The Avonian succession in the south of the Isle of Man. *Q. J. geol. Soc. Lond.*, **86**, 234–90.

MAGRAW, D., CLARKE, A. M. and SMITH, D. B. 1963. The stratigraphy and structure of part of the south-east Durham coalfield. *Proc. Yorks. geol. Soc.*, **34**, 153–208.

MILLER, A. A. and TURNER, J. S., 1931. The Lower Carboniferous succession along the Dent Fault and the Yoredale Beds of the Shap district. *Proc. Geol. Ass.*, **42**, 1–28.

MILLS, D. A. C. and HULL, J. H. 1968. The Geological Survey Borehole at Woodland, Co. Durham (1962). *Bull. geol. Surv. Gt Br.*, No. 28, 1–37.

MOSELEY, F. 1953. The Namurian of the Lancaster Fells. *Q. J. geol. Soc. Lond.*, **109**, 423–54.

—— and AHMED, S. W. 1967. Carboniferous joints in the north of England and their relationship to earlier and later structures. *Proc. Yorks. geol. Soc.*, **36**, 61–90.

OWENS, B. and BURGESS, I. C. 1965. The stratigraphy and palynology of the Upper Carboniferous outlier of Stainmore, Westmorland. *Bull. geol. Surv. Gt Br.*, No. 23, 17–44.

RAISTRICK, A. 1934. The correlation of coal seams by microspore content. *Trans. Inst. Min. Engrs.*, **88**, 142–53.

RAYNER, D. H. 1953. The Lower Carboniferous rocks in the north of England: a review. *Proc. Yorks. geol. Soc.*, **28**, 231–315.

RICHARDSON, G. 1965. In *Summ. Prog. geol. Surv. Gt Br. for 1964*, 59.

ROBSON, D. A. 1956. A sedimentary study of the Fell Sandstones of the Coquet valley, Northumberland. *Q. J. geol. Soc. Lond.*, **112**, 241–62.

ROWLEY, C. R. 1969. The stratigraphy of the Carboniferous Middle Limestone Group of west Edenside, Westmorland. *Proc. Yorks. geol. Soc.*, **37**, 329–50.

SIMPSON, J. B. 1902. The probability of finding workable seams of coal in the Carboniferous Limestone or Bernician Formation, beneath the regular Coal-Measures of Northumberland and Durham, with an account of a recent deep boring in Chopwell Woods, below the Brockwell seam. *Trans. Instn Min. Engrs.*, **25**, 549–71.

SMITH B. 1927. On the Carboniferous Limestone Series of the northern part of the Isle of Man. In *Summ. Prog. geol. Surv. Gt Br. for 1926*, 108–19.

SMITH, S. 1912. The Carboniferous Limestone formation of the north of England. *North of Engl. Instn of Min. Mech. Engrs.* Newcastle upon Tyne, ix+231 pp.

SMITH, T. E. 1967. A preliminary study of the sandstone sedimentation in the Lower Carboniferous of the Tweed Basin. *Scot. J. Geol.*, **3**, 282–305.

TAYLOR, B. J. 1961. The stratigraphy of exploratory boreholes in the west Cumberland coalfield. *Bull geol. Surv. Gt Br.*, No. 17, 1–74.

TROTTER, F. M. 1951. Sedimentation facies in the Namurian of north-western England and adjoining areas. *Lpool Manchr geol. J.*, **1**, 77–112.

TURNER, J. S. 1927. The Lower Carboniferous succession in the Westmorland Pennines and the relations of the Pennine and Dent faults. *Proc. Geol. Ass.*, **38**, 339–74.

WESTOLL, T. S., ROBSON, D. A. and GREEN, R. 1955. A guide to the geology of the district around Alnwick, Northumberland. *Proc. Yorks. geol. Soc.*, **30**, 61–100.

WOOLACOTT, D. 1923. A boring at Roddymoor Colliery, near Crook, Co. Durham. *Geol. Mag.*, **60**, 50–62.

Intrusive Igneous Rocks

BROWN, P. E., MILLER, J. A. and SOPER, N. J. 1964. Age of the principal intrusions of the Lake District. *Proc. Yorks. geol. Soc.*, **34**, 331–42.

DUNHAM, A. C. and KAYE, M. J. 1965. The petrology of the Little Whin Sill, County Durham. *Proc. Yorks. geol. Soc.*, **35**, 229–76.

DUNHAM, K. C., DUNHAM, A. C., HODGE, B. L. and JOHNSON, G. A. L. 1965. Granite beneath Viséan sediments with mineralization at Rookhope, northern Pennines. *Q. J. geol. Soc. Lond.*, **121**, 383–417.

DWERRYHOUSE, A. R. 1909. Intrusive rocks in the neighbourhood of Eskdale. *Q. J. geol. Soc. Lond.*, **65**, 55–80.

FIRMAN, R. J. 1957. Fissure metasomatism in volcanic rocks adjacent to the Shap Granite. *Q. J. geol. Soc. Lond.*, **113**, 205–22.

FITCH, F. J. and MILLER, J. A. 1967. The age of the Whin Sill. *Geol. J.*, **5**, 233–50.

GRANTHAM, D. R. 1928. The petrology of the Shap Granite. *Proc. Geol. Ass.*, **39**, 299–331.

HARKER, A. 1892. The lamprophyre dykes of the north of England. *Geol Mag.*, **9**, 199.
—— 1894. Carrock Fell. A study in the variation of igneous rock masses. Pt I. The Gabbro. *Q. J. geol. Soc. Lond.*, **50**, 311–37.
—— 1895. Carrock Fell. A study in the variation of igneous rock masses. Pt II. The Carrock Fell Granophyre. Pt III. The Grainsgill greisen. *Q. J. geol. Soc. Lond.*, **51**, 125–48.
—— 1902. Notes on the igneous rocks of the English Lake District. *Proc. Yorks. geol. Soc.*, **14**, 487–96.
—— and MARR, J. E. 1891. The Shap Granite and associated rocks. *Q. J. geol. Soc. Lond.*, **47**, 266–328.

HITCHEN, C. S. 1934. The Skiddaw Granite and its residual products. *Q. J. geol. Soc. Lond.*, **90**, 158–99.

HOLLAND, J. G. 1967. Rapid analysis of the Weardale Granite. *Proc. Yorks. geol. Soc.*, **36**, 91–113.

HOLMES, A. and HARWOOD, H. F. 1928. The age and composition of the Whin Sill and the related dykes of the north of England. *Mineralog. Mag.*, **21**, 493–524.
—— —— 1929. The tholeiitic dykes of the north of England. *Mineralog. Mag.*, **22**, 1–52.

HORNUNG, G. AL-ANI A. and STEWART, R. M. 1966. Composition and emplacement of the Cleveland Dyke. *Trans. Leeds geol. Ass.*, **7**, 232–49.

HUDSON, S. N. 1937. The volcanic rocks and minor intrusions of the Cross Fell Inlier, Cumberland and Westmorland. *Q. J. geol. Soc. Lond.*, **93**, 368–405.

RASTALL, R. H. 1906. The Buttermere and Ennerdale granophyre. *Q. J. geol. Soc. Lond.*, **62**, 253–73.
—— 1910. The Skiddaw Granite and its metamorphism. *Q. J. geol. Soc. Lond.*, **66**, 116–41.

SIMPSON, A. 1964. Deformed acid intrusions in the Manx Slate Series, Isle of Man. *Geol. J.*, **4**, 189–206.
—— 1965. The syntectonic Foxdale-Archallagan granite and its metamorphic aureole, Isle of Man. *Geol. J.*, **4**, 415–34.

SPEARS, D. A. 1961. Joints in the Whin Sill and associated sediments in Upper Teesdale, northern Pennines. *Proc. Yorks. geol. Soc.*, **33**, 21–30.

TURNER, J. S. 1935. Structural geology of Stainmore. *Proc. Geol. Ass.*, **46**, 121–51.

Permian and Triassic

ARTHURTON, R. S. 1971. The Permian evaporites of the Langwathby Bore, Vale of Eden, north-west England. *Rep. Inst. geol. Sci. Gt Br.* (*In press.*)

—— and HEMINGWAY, J. E. In the press. The St. Bees Evaporites: a carbonate-evaporite formation of Upper Permian age in west Cumberland, England. *Proc. Yorks. geol. Soc.*, **38**.

BURGESS, I. C. 1965. The Permo-Triassic rocks around Kirkby Stephen, Westmorland. *Proc. Yorks. geol. Soc.*, **35**, 91–101.

DUNHAM, K. C. and ROSE, W. C. C. 1948. Permo-Triassic geology of south Cumberland and Furness. *Proc. Geol. Ass.*, **60**, 11–40.

GOODCHILD, J. G. 1892. Observations on the New Red Series of Cumberland and Westmorland, with especial reference to classification. *Trans. Cumb. Westm. Ass.*, **17**, 1–24.

HOLLINGWORTH, S. E. 1942. Correlation of Gypsum-Anhydrite deposits in the North of England. *Proc. Geol. Ass.*, **53**, 141–51.

SHERLOCK, R. L. and HOLLINGWORTH, S. E. 1938. Gypsum and anhydrite; celestine and strontianite. 3rd Edit. *Mem. geol. Surv. Gt Br. Min. Resources*, **3**, v+94 pp., H.M.S.O., London.

SMITH, B. 1924. On the west Cumberland brockram and associated rocks. *Geol. Mag.*, **61**, 289–308.

SMITH, D. B. 1970. The Permian and Trias in *The geology of Durham County*, G. A. L. Johnson, Ed. *Trans. nat. Hist. Soc. Northumb.*, **41**, 66–91.

STEWART, F. H. 1954. Permian evaporites and associated rocks in Texas and New Mexico compared with those of northern England. *Proc. Yorks. geol. Soc.*, **29**, 185–235.

TRECHMANN, C. T. 1913. On a mass of anhydrite in the Magnesian Limestone at Hartlepool, and on the Permian of south-eastern Durham. *Q. J. geol. Soc. Lond.*, **69**, 184–218.

—— 1914. On the lithology and composition of Durham Magnesian Limestone. *Q. J. geol. Soc. Lond.*, **70**, 232–65.

—— 1925. The Permian Formation in Durham. *Proc. Geol. Ass.*, **42**, 246–52.

WOOLACOTT, D. 1919. The Magnesian Limestone of Durham. *Geol. Mag.*, **6**, 452–65, 485–98.

Quaternary

ANDERSON, W. 1940. Buried valleys and Late-glacial drainage systems in North-west Durham. *Proc. Geol. Ass.*, **51**, 274–81.

BEAUMONT, P. 1968. A history of Glacial research in northern England from 1860 to the present day. *Univ. Durham Occ. Papers*, No. 9.

GODWIN, H., WALKER, D. and WILLIS, E. H. 1957. Radiocarbon dating and post-glacial vegetational history: Scaleby Moss. *Proc. Roy. Soc.* (B), **147**, 352–66.

GRESSWELL, R. K. 1951. The glacial geomorphology of the south-eastern part of the Lake District. *Lpool Manchr geol. J.*, **1**, 57–70.

HOLLINGWORTH, S. E. 1931. The glaciation of western Edenside and adjoining areas, and the drumlins of Edenside and the Solway Basin. *Q. J. geol. Soc. Lond.*, **87**, 281–359.

MANLEY, G. 1959. The Late-glacial climate of north-west England. *Lpool Manchr geol. J.*, **2**, 188–215.

MITCHELL, G. F. 1965. The Quaternary deposits of the Ballaugh and Kirkmichael districts, Isle of Man. *Q. J. geol. Soc. Lond.*, **121**, 359–81.

PENNY, L. F. 1964. A review of the last glaciation in Great Britain. *Proc. Yorks. geol. Soc.*, **34**, 387–411.

RAISTRICK, A. 1931. The glaciation of Northumberland and Durham. *Proc. Geol. Ass.*, **42**, 281–91.

—— 1934. The correlation of glacial retreat stages across the Pennines. *Proc. Yorks. geol. Soc.*, **22**, 199–214.

TRECHMANN, C. T. 1915. The Scandinavian Drift of the Durham coast and the general glaciology of south-east Durham. *Q. J. geol. Soc. Lond.*, **71**, 53–82.

TROTTER, F. M. 1929. The glaciation of eastern Edenside, the Alston Block and the Carlisle Plain. *Q. J. geol. Soc. Lond.*, **85**, 558–612.

WALKER, D. 1955. Studies in the post-glacial history of British vegetation. IV. Skelsmergh Tarn and Kentmere, Westmorland. *New. Phytol.*, **54**, 222–54.

WOOLACOTT, D. 1921. The interglacial problem and the glacial and post glacial sequence in Northumberland and Durham. *Geol. Mag.*, **58**, 64–74.

Geophysical Investigations

BOTT, M. H. P. 1964. Gravity measurements in the north-eastern part of the Irish Sea. *Q. J. geol. Soc. Lond.*, **120**, 369–396.

—— 1967. Geophysical investigations of the northern Pennine basement rocks. *Proc. Yorks. geol. Soc.*, **36**, 139–68.

—— and MASSON SMITH, D. J. 1957. The geological interpretation of a gravity survey of the Alston Block and the Durham coalfield. *Q. J. geol. Soc. Lond.*, **113**, 93–117.

CLARKE, A. M., CHAMBERS, R. E., ALLONBY, R. H. and MAGRAW, D. 1961. A marine geophysical survey of the undersea coalfields of Northumberland, Cumberland and Durham. *Min. Engr.*, No. 15, 197–215.

HOSPERO, J. and WILLMORE, P. L. 1953. Gravity measurements in Durham and Northumberland. *Geol. Mag.*, **90**, 117–26.

POOLE, G., WHETTON, J. T. and TAYLOR, A. 1935. Magnetic observations on concealed dykes and other intrusions in the Northumberland coalfield. *Trans. Instn Min. Engrs.*, **89**, 34–47.

ROBSON, D. A. 1964. The Acklington Dyke—a Proton Magnetometer survey. *Proc. Yorks. geol. Soc.*, **34**, 293–308.

Economic Geology

CARRUTHERS, R. G. and ANDERSON, W. 1943. Some refractory materials in north-eastern England. *Wartime Pamphlet, geol. Surv. Gt Br.*, No. 31.

DAVIES, W. and REES, W. J. 1943. British resources of steel moulding sands. *Iron & Steel Inst.*, **148**, 11–111.

DUNHAM, K. C. 1944a. The genesis of the north Pennine ore deposits. *Q. J. geol. Soc. Lond.*, **90**, 689–720.

—— 1944b. The production of galena and associated minerals in the northern Pennines; with comparative statistics for Great Britain. *Trans. Instn Min. Metall.*, **53**, 181–252.

—— 1948. Geology of the Northern Pennine Orefield: Vol. 1, Tyne to Stainmore. *Mem. geol. Surv. Gt Br.*, vi+367 pp.

—— 1959a. Epigenetic mineralization in Yorkshire. *Proc. Yorks. geol. Soc.*, **32**, 1–29.

—— 1959b. Non ferrous mining potentialities of the northern Pennines: pp. 115-47 in *The future of non-ferrous mining in Great Britain and Ireland.* xxvi+614. Instn Min. Metall., London.

—— 1960. Syngenetic and diagenetic mineralization in Yorkshire. *Proc. Yorks. geol. Soc.*, **32**, 229–84.

—— 1967. Mineralization in relation to the pre-Carboniferous basement rocks, northern England. *Proc. Yorks. geol. Soc.*, **36**, 195–201.

—— and DINES, H. G. 1945. Barium Minerals in England and Wales. *Wartime Pamphlet geol. Surv. Gt Br.*, No. 46.

—— and ROSE, W. C. C. 1941. Geology of the iron-ore field of south Cumberland and Furness. *Wartime Pamphlet, geol. Surv. Gt Br.*, No. 16.

EASTWOOD, T. 1959. The Lake District mining field: pp. 149–74 in *The future of non-ferrous mining in Great Britain and Ireland.* xxvi+614. Instn Min. Metall., London.

HOPKINS, W. 1954. The coalfields of Northumberland and Durham: pp. 289–313 in Trueman, A. E. *The Coalfields of Great Britain.* Edward Arnold, London.

POSTLETHWAITE, J. 1913. Mines and Mining in the (English) Lake District (Second Edition). vii+164. Moss and Sons, Whitehaven.

TROTTER, F. M. 1953. The Cumberland Coalfield: pp. 314 in Trueman, A. E., *The Coalfields of Great Britain.* Edward Arnold, London.

The following Special Reports on the Mineral Resources of Great Britain contain references to Northern England. These are Memoirs of the Geological Survey of Great Britain, published by Her Majesty's Stationery Office, London. All are out of print.

Vol.	I	1923	*Tungsten and Manganese Ores* (third edition)
Vol.	II	1922	*Barytes and Witherite* (third edition)
Vol.	III	1938	*Gypsum and Anhydrite* (third edition)
Vol.	IV	1952	*Fluorspar* (fourth edition)
Vol.	V	1917	*Potash-Felspar, Phosphate of Lime, Alum Shales, Plumbago or Graphite, Molybdenite, Chromite, Talc and Steatite (Soapstone, etc.), Diatomite* (second edition)
Vol.	VI	1920	*Ganister and Silica Rock* (second edition)
Vol.	VIII	1924	*Iron Ores: Haematites of West Cumberland, Lancashire and the Lake District* (second edition)
Vol.	IX	1919	*Iron Ores: Sundry unbedded Ores of Durham, East Cumberland, etc.*
Vol.	XIV	1920	*Refractory materials: Fireclays*
Vol.	XVIII	1921	*Rock-Salt and Brine*
Vol.	XXII	1921	*Lead and Zinc Ores of the Lake District*
Vol.	XXV	1923	*Lead and Zinc Ores of Northumberland and Alston Moor.*
Vol.	XXVI	1923	*Lead and Zinc Ores of Durham, Yorkshire and Derbyshire, with notes on the Isle of Man.*
Vol.	XXX	1925	*Copper Ores of the Midlands, Wales, the Lake District and the Isle of Man.*

Index

Printed in England for Her Majesty's Stationery Office
by Hull Printers Limited, Willerby, Hull HU10 6DH

Dd. 502165   K160.